BLACK GOLD

The Story of Texas Oil & Gas

by Dr. Roger Olien

Published for the Texas Energy Museum and The Petroleum Museum

Historical Publishing Network
A division of Lammert Incorporated
San Antonio, Texas

Black Gold: The Story of Texas Oil & Gas

author:	Roger Olien
cover artist:	H. Phenix
contributing writers for "Sharing the Heritage":	Judi Free
	Marie Beth Jones
	Scott Williams

Historical Publishing Network

president:	Ron Lammert
vice president:	Barry Black
project managers:	Curtis Courtney
	Barbara Lane
	Sydney McNew
	Lou Ann Murphy
	Joe Neely
	Robin Neely
	Roger Smith
	Robert Steidle
director of operations:	Charles A. Newton, III
administration:	Angela Lake
	Donna M. Mata
	Judi Free
book sales:	Dee Steidle
graphic production:	Colin Hart
	Michael Reaves
	Craig Mitchell
	John Barr
	Desirie Vargas

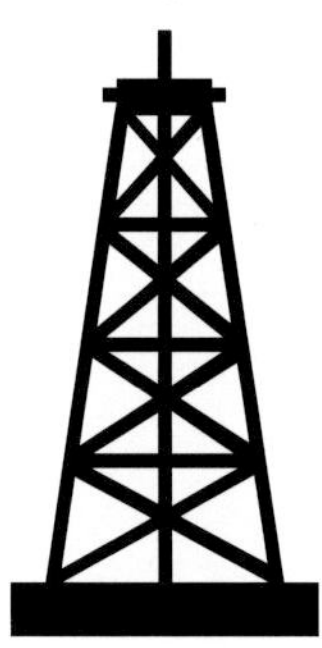

Contents

FOREWORD

A word about names and terms: The histories of company names are sometimes even more complicated than family genealogies, mainly because of acquisitions and mergers. In this book, I have generally used the names most familiar to readers, even when the companies were not called by their most modern names when they appear in the story. For example, Magnolia Oil and Refining became a subsidiary of Socony-Vacuum, which was the Standard Oil Company of New York plus the Vacuum Oil company. All of these companies are referred to as Mobil. Similarly, the Texas Company and its predecessors are identified as Texaco. The only general exception is that Humble Oil and Refining, now part of Exxon/Mobil, appears as Humble.

I will refer to geographical and geological areas in general terms with only occasional reference to their technical identifications. Thus, the upper Gulf Coast refers to the part of Texas that lies from the vicinity of Houston to the Louisiana border. Similarly, I will often include the plains north of Midland and Odessa in the Permian Basin even though the areas are geographically distinct because Midland emerged as the management center during the 1920s, as did Odessa in the service and supply arms of the industry. Geological formations are usually referred to in more general terms than geoscientists use because more accurate technical language would make the story more difficult for most readers to follow. Similarly, drilling, production, refining, and petrochemicals are described with a minimum of technical language. Readers who are interested in more details of this kind should consult the books that are listed at the end of this one.

By way of thanks, I want to recognize the hundreds of men and women who have shared their experiences in the Texas oil industry with me. Among them, J. Conrad Dunagan, Kelly Bell, and John Redfern did their best to help me understand the complexity, dynamics, and excitement of the search for black gold. Jack Porter and John Tower expanded my understanding of the relationship between the politics of oil and the Texas Republican Party. A special debt is owned to the University of Texas of the Permian Basin. Under the leadership of President David Watts and his predecessors, UTPB paid me to do what I enjoy most—teach and write history—for more than three decades. I have shared the concept that led me to write about Texas oil again with a number of people, notably Jane Phares, Ryan Smith, and Ron Lammert. Responsibility for the specific content of this book—including any incidental errors that might appear in it—are mine alone.

Roger M. Olien
Austin, Texas
2004

The trading center of Old Time Texas, the

Galveston Cotton Exchange.

INTRODUCTION

A PROMISING BEGINNING

Rice harvesters at work on the Gulf Coast.
COURTESY OF THE HOUSTON METROPOLITAN
RESEARCH CENTER, HOUSTON PUBLIC LIBRARY.

"What, in this age when we are all so oversupplied with information, does a given human being need to remember other than, perhaps, the names of his or her spouse (if any) and children?"
- Larry McMurtry, *Walter Benjamin and the Dairy Queen*

"As I grow older, I value more highly the fellow who does things. He is so rare. And he is so often more valuable to society than many more intelligent, but less energetic fellows."
- Wallace E. Pratt to Everett DeGolyer, June 21, 1956

McMurty poses a fair question. How much of the past is so important now that it is still remembered—that anybody should be expected to remember it in the future? There is no quick and easy answer to this question, as to most truly important ones, but as far as the story of oil in Texas is concerned, I will take Wallace Pratt's perspective and write about people, placing a few essential topics at the top of the list: significant individuals and groups—wildcatters, workers, corporate executives, political leaders—the Texans who shaped the petroleum industry; the impacts their work have had on lives of Texans; the inter-action of business and politics in Texas, where politics is still a uniquely variable and dynamic combination of three-card monte, carnival, and bull-riding.

The telling of the story will include most parts of Texas and most groups within it, though the American oil industry began far from the coastal lowlands, piney woods, and semiarid plains where it has been for more than a century.

The American petroleum industry began its dramatic history, far from Texas, in the backwoods of Pennsylvania on the eve of the Civil War. Thereafter, oilmen brought in gushers that spewed thousands of barrels a day in those oil regions, and successively, in West Virginia, New York, Ohio, Indiana, Illinois, California, and Kansas before Texans produced enough crude oil to leave a big spot on the dirt roads that joined its thousands of small, rural towns. There was a small start near Nacogdoches during the 1860s, but the experiment was slow to unfold into the 1890s and the results were of little importance outside the immediate area.

Texans still relied on the surface of the land for as much of a living as they could scratch out of it during the 1890s. From Wichita Falls to Houston, "eating regular" depended on the highly volatile cotton prices. As farmers prospered and failed, riding wild swings in prices, hanging on for dear life, like passengers on amusement park roller coasters, occasional fat years were followed regularly by lean ones, as more and more Texans plunged into tenant farming and poverty.

At the beginning of the twentieth century, the lumber industry was the state's largest industrial employer. At that, it provided jobs for no more than ten thousand workers, many of them,

Symbols of Old Time Texas: Young West Texas cowboys looking tough for the camera.
COURTESY OF THE PERMIAN BASIN PETROLEUM MUSEUM, LIBRARY, AND HALL OF FAME, MIDLAND, TEXAS.

black and white alike, displaced from large families on marginal cotton land. The work was hard, though it paid them better to hew trees, drive animals and feed the saw mills than plowing, planting, chopping and picking cotton had ever done. About the same number of farmboys hauled crops and breathed dust and lint, processing cotton at gins and compresses. Their fortunes, too, like those of their employers, rose and fell with cotton prices, and the weather.

By 1900 there was one small encouraging sign on the horizon because, near the end of the nineteenth century, another oil discovery, this time at Corsicana, put Texas on the national oil map for the first time. Modest enough by the standards of the older producing regions, nearly three thousand Texans, native and immigrant, earned a living in the new industry. Among 2.5 million Texans, their numbers were modest—statistically insignificant in the whole economy of the state—there were more than 650,000 Texans in agribusiness, but these oilmen and workers were the cutting edge of economic diversification; the pioneers in an industry that would sweep across all sections of the state, bringing widespread social and economic change during the next half century.

The Corsicana story began when city fathers looked at the prospects of that county seat town, dependent on cotton grown on land less fertile than that of East Texas counties, and decided that its future prosperity depended on their ability to attract new industry—companies not dependent on agriculture. After a preliminary survey of the town's relative advantages, they concluded that its central location and railroad connections with both Dallas and Houston might well be attractive to new employers if they were able to supplement the barely adequate water supply with new wells. Using local funds, community leaders contracted the drilling of new water wells during 1894, with a surprising and not totally pleasing result, the driller found oil. There was enough oil to create problems with new water wells, from 2.5 to 25 barrels of oil per day, but, initially, not enough to launch a new industry. So, they tried again and again with

The reality of Old Time Texas: Loading cotton barges on Buffalo Bayou, Houston.
COURTESY OF THE HOUSTON METROPOLITAN RESEARCH CENTER, HOUSTON PUBLIC LIBRARY.

the same result to the point that the likelihood of producing enough oil to justify exploration for it increased with every polluted water well.

And as early wells were completed during the early months of 1897, production reached 180 barrels per day, and Texas was on the edge of its first oil boom. Corsicana shopkeepers, lawyers, bankers, politicians, virtually everybody with spare change, jumped into the game, buying leases, forming partnerships, laying primitive pipelines, and producing enough crude by the end of 1897 to turn their minds and ambitions to the creation of a new industry. Some of Corsicana's new oil men, including James Autry, James A. Garrity, Walter B. Sharp, and Walter W. Fondren, launched careers that would in time make them giants in the new Texas oil elite. Nor were they alone in cashing in on the new oil "play." Farm boys, hired on

as roughnecks and roustabouts, were doing unskilled and semiskilled work on rig floors and elsewhere in the field. The more fortunate and enterprising blue collar Texans learned how to drill wells and operate refineries, skills that would boost their fortunes in years to come. Beginning in Corsicana, entrepreneurs and blue collar workers alike, Texans staked their livelihoods on the new industry.

Lacking experience with the technical aspects of oil, especially refining and distribution, community leaders succeeded in luring Joseph Stephen Cullinan, a modestly successful oilman from the East, to Texas. Cullinan brought the "know-how" needed to create and operate most of the phases of the multiphase industry including exploration, pipelines, and refining. While he was in Corsicana, he also led the new venture into sales, vending fuel oil to railroads and sludgy asphalt to towns that upgraded their dirt streets. Above all, he created a market for crude oil, encouraging more exploration and production, which reached 2,275 barrels a day in 1900.

As important as the industry was in Corsicana, it had scant impacts in the rest of the state and nation. It was, however, more important than employment and production data indicate because it got Texans thinking about oil and the possibilities of finding even more of it in other parts of the Lone Star State.

CHAPTER I

A MIGHTY ROAR—AND TEXAS HAS OIL!

"The world stands in wonder..."
- *The Beaumont Enterprise*, 1901

The lure of oil caught the imagination of Patillo Higgins, a brainy eccentric, who reasoned that the seepage of natural gas around a salt dome south of Beaumont was a likely indication of the presence of enough crude oil to drill for. It took considerable imagination to reach that conclusion because seeps had never been, nor are they now, reliable evidence of the presence of commercial quantities of petroleum. Undaunted by discouragement by geologists and local businessmen, Higgins stuck to his opinion with dogged tenacity—in most situations one of his less endearing characteristics. His fantasy—for that's what it was unless you believe in ESP—convinced several wealthy Beaumont capitalists to ante up for Higgins' first wildcat well. He promised lumber magnate George Washington Carroll, "I will make millions for both of us." The effort, drilled by Walter B. Sharp from Corsicana, ran into driving rains and high winds. It was slow going. The wildcat test took more time and money than Higgins could raise so he abandoned it in 1893.

Workers hauling timbers to the new oil field near Beaumont.
COURTESY OF THE TEXAS ENERGY MUSEUM.

With the blind conviction of a true believer, Higgins persisted and raised funds for a second test. This time poor equipment and inexperienced drillers brought the venture to the same end as the first. The well was plugged and abandoned. So far, it was two strikes against Higgins in the oil game; that was as far as he went, for the time being. The third test proceeded over Higgins' objections as to the qualifications of the driller—the same man who had botched the second try—so Higgins opted out of the new company accepting a small settlement for his part-ownership. Thereafter, he watched from the sidelines as Act Three of the drama continued with the abandonment of another well, much as Higgins had predicted. After three unsuccessful tries, it would have been reasonable for Higgins to turn his attention back to his brick factory and write off oil. However, as even his friends conceded, Higgins was not by nature reasonable.

No longer credible among local backers, Higgins enlisted science to support his belief that the salt dome, later called Spindletop locally, would be the scene of an oil bonanza. His expert, Anthony Lucas, a Montenegrin naval engineer once in the service of Austria, had acquired experience with salt domes drilling for salt in Louisiana. He knew just enough about the geological formation to believe that a vertical salt plug could hold oil in horizontal geological formations. With that possibility in mind, Lucas met with Higgins, examined the area around Spindletop, and organized a drilling operation, without Higgins as an investor, a development led to litigation down the road.

Lucas set up operations in Beaumont, but his results were no better than Higgins'. After coping with quicksand, which swallowed the drill bit, Lucas hit a pocket of high pressure natural gas, which destroyed the rig, and most of his investment. Nearly as stubborn as Higgins, Lucas persisted, probably, because the failed wildcat had found sand streaked with crude oil—a sign of the probable presence of oil. When Lucas tried to raise money in Beaumont for another test, he found that local investors had heard talk of black gold before, mainly from Higgins, so Lucas

turned to James M. Guffey and John H. Galey, Pennsylvania oilmen. They convinced Pittsburgh capitalist Andrew Mellon to provide funds for three more tests carried out by a new company that conspicuously omitted Patillo Higgins.

Back in Beaumont, Lucas leased the equivalent of twenty-three sections of land in the vicinity of the fourth test, planned for twelve hundred feet, twice the depth reached by earlier drillers. Acting on advice from John Galey, he hired Jim, Al, and Curt Hamill, three strong-jawed brothers who had drilled successfully at Corsicana. He could not have done better. The Texans combined Patillo Higgins' tenacity with practical experience and the ability to improvise solutions to problems as they encountered them.

On the scene in Beaumont, the Hamills built a rig out of green timber, hauled in a boiler and other equipment, and cut into the earth with a sharp twelve-inch bit on October 27, 1900. They progressed without serious complications until they hit the quicksands that had bedeviled earlier drillers at 160 feet. They were still more than a thousand feet short of their objective and likely to get no farther unless they could hold the sand back and drill through to harder mineral formations. After some discussion, the brothers pounded eight-inch pipe through the sand. Then they ran the four-inch drill pipe down through it, "casing" the well, as the technique would come to be called in the oil industry.

As they continued to drill, the Hamills encountered high pressure gas pockets, the kind that had destroyed an earlier try, but they improvised another solution; mixing clay with the water they encountered, producing a heavy kind of mud that held the gas back and permitted them to drill ahead. The circulating "mud", as it would commonly be called in the oil industry, lessened the gas problem and brought up increasingly impressive shows of crude oil as the drilling proceeded. At nine hundred feet, the Hamills, Lucas, and Galey were optimistic, but much less so as further drilling disclosed little more than oil-streaked sand. Reluctant to throw good money after bad, Galey proposed abandoning the test, but agreed to continue after Al Hamill and Caroline Lucas urged him to drill three hundred feet deeper. The Hamills drilled on, with little encouragement from Beaumonters, who had gone through the emotional rollercoaster of hope and disappointment with earlier wells. They worked until Christmas Eve of 1900, when

Above: A forest of derricks on the tract developed by former Governor James Hogg and his associates.

COURTESY OF THE TEXAS ENERGY MUSEUM ARCHIVES.

Below: Fire was a constant danger at the new oil field.

COURTESY OF THE TEXAS ENERGY MUSEUM ARCHIVES.

they shut down and went home to Corsicana for the holidays, returning on New Year's Day of the new century discouraged by the slow pace of the project.

Once they were back on the job, however, drilling went somewhat faster, through 140 feet of limestone, sulphur and other minerals, until they hit harder rock where their drilling pipe jammed, wearing out the drill bit. Now more pessimistic, the brothers made repairs and began to drill again on January 10. On the morning of that historic day, they refitted with a new bit as a stiff north wind blasted the rig, and went back to making hole. Before the brothers reached their target depth, the well blew itself in. First, the rig shook and groaned then drilling mud surged out of the hole and over the rig floor. Before the brothers could share their common-sense reaction— they had seen blow-outs before—the dramatic show by nature continued as flowing oil, tons of mud, water and rock shot out of the hole and tore through the top of the derrick. The Hamills sped out of harm's way. From a safe distance, they watched the drilling pipe rise out of the well, flying hundreds of feet into the air. Then, there was quiet, ominous silence because their rig was in ruins and they still had not produced a barrel of oil.

As they prepared to clean up the wreck, the well thundered again, throwing out more

mud, followed by natural gas—another familiar and discouraging sequence. When they went to the wellhead to investigate, they heard a loud gurgling sound as more liquid rose. A few seconds later oil flowed slowly, and then gushed over the top of the battered rig and over the surrounding countryside. Anthony Lucas and the Hamill brothers had their well, and they had made history!

For several days, the oilmen, Beaumont residents, and a flood of visitors were too excited to worry about history. Some speculated on the origins of the oil, guessing that there was a river of oil that flowed southeast from Corsicana, through Beaumont, and out to sea where slicks were commonly observed. Whatever its source, the oil was undeniably real as it continued to flow and spray, covering Beaumont houses with a corrosive dark film. Would-be investors and speculators crowded into trains that arrived from Houston, rushed to buy leases, sell shares in new companies, and crowd into hotels and dining rooms. As was said at the time, every man and his dog wanted to get into the oil business. The *Beaumont Enterprise* set up a sideline printing oil lease forms and sold copies of the famous gusher photo by

Frank J. Trost. Everybody wanted to share the excitement and profit from it. Four days after the well roared in, a local dyer and cleaner advertised "Another Well Discovered."

> While the oil well is sprouting
> And the Beaumont folks are shouting
> And Lucas has realized his dream—
> Just remember I'm still working

Above: As close together as they could drill them: Boiler Avenue at Spindletop.
COURTESY OF THE TEXAS ENERGY MUSEUM.

Below: Roustabouts pose at a Guffey-Galey Petroleum Company (Gulf Oil) rig at Spindletop, 1903.
COURTESY OF THE C. C. RISTER COLLECTION, SOUTHWEST COLLECTION, TEXAS TECH UNIVERSITY.

Above: *Curbstone oil traders trading leases and speculating on company stock and oil prices in Beaumont, 1901. This photo originally appeared in the* Oil and Gas Journal.

COURTESY OF THE C. C. RISTER COLLECTION, SOUTHWEST COLLECTION, TEXAS TECH UNIVERSITY.

Below: *The Guffey-Galey (Gulf) Refinery at Port Arthur, the modest beginnings of the vast refining operations on the Gulf Coast.*

COURTESY OF THE C. C. RISTER COLLECTION, SOUTHWEST COLLECTION, TEXAS TECH UNIVERSITY.

And my business I'm not shirking
So bring me all of your soil clothes to clean.

The Spindletop gusher was the stuff that legends are made of. Versions of the Spindletop story found their way into poems, melodramas, dime novels and songs. One of the latter, the *Lucas Geyser*, a march with words, captured the spirit of the occasion: "You talk about your Klondike rush and gold in frozen soil, But it don't compare with Beaumont rush when Lucas he struck oil." Fortunately, the flow of Spindletop oil was stronger than the inspiration of these poets and musicians.

Patillo Higgins claimed credit for the discovery and his descendants still defend his assertions even though geoscientists have generally sided with Anthony Lucas. Writing half a century later, Wallace Pratt, Humble's famed geologist, applauded Lucas:

He deserved credit, I thought for recognizing the salt dome character of Spindletop. But other contemporary students of Gulf Coast geology also suspect the eminence of Spindletop to overlie a salt dome. What Lucas really deserved credit for was the courage of his conviction. While others talked about Spindletop and a possible salt dome, Lucas did something about it. He drilled a well and proved it. Incidentally, he discovered a great oil field.

As days passed, it became clear to excited observers that the discovery was more than an incident. The well flowed 250,000 barrels in a few more days, qualifying as the biggest gusher so far in the United States. Texas was ushered into its Oil Age by a European immigrant and his determined wife, an old time wildcatter from Pennsylvania, three central Texas farm boys, and the African Americans who had hauled pipe to the rig and cooked for the crew. Where ever they came from and whatever their races and ethnic origins, they were pure Texas and they give place only to the generation of Sam Houston in writing Lone Star history with their deeds.

Had Spindletop been an isolated discovery, it would have been of no more than passing significance. Guffey and Galey had seen one-well oil fields before, but as oilmen followed up that discovery by drilling on every visible salt dome, they found even more oil In quick succession, they brought in prolific fields at Sour Lake, Saratoga, Batson, Humble, and Goose Creek. The race was on. At Sour Creek, drillers completed 450 wells by the

end of 1903. These wells typically paid for themselves in a few weeks because they came in at about 450 feet. The new fields added capital and experience to the growing industry. Saratoga and Batson were developed in 1904 with Humble following the next year.

Once wildcatters discovered oil, they drilled as quickly as they could and they sited rigs so close together that workers claimed that they could jump from one to the other, crossing the fields without letting their shoes hit the ground. A large part of the motivation to drill quickly and close came from current legal interpretations of property rights. In Texas, as in other producing states, it was held that oil belonged to the person who produced it even though his well might be draining crude from under adjoining leases. The "rule of capture," as it was called, typically produced the rush to drill; that, in turn, created oil booms. The familiar photographs of the forests of derricks in oil fields reflects both the legal and the operating situations that oilmen coped with for decades.

Exploration spread along Galveston Bay in 1903, but the hopeful wildcatters did not locate producible oil. Five years later, Houston oilmen drilled in the marsh at Goose Creek and struck oil at sixteen hundred feet. Their success was short lived. In the same field, Texaco drilled twenty dry holes in succession, put a damper on interest in the

new field, and slowing exploration in it. Local pessimists revised their opinions in 1916, when Charles Mitchell brought in a ten-thousand-barrel gusher on shore, triggering what was, by then, the familiar rush of operators, drillers, suppliers, and blue collar workers to the area. Two years later, the field produced about nine million barrels of oil, and oilmen moved offshore drilling from piers they built along the creek and into adjoining bays—producing the first offshore oil in Texas.

The steady flow of oil from successive discoveries kept the oil industry booming in the upper Gulf Coast area. Although production in most of the salt dome

Above: The first oil field "greasy spoon" restaurants, under tents, at Beaumont, 1902.
COURTESY OF THE C. C. RISTER COLLECTION PHOTO, SOUTHWEST COLLECTION, TEXAS TECH UNIVERSITY.

Below: Oil terminals at Port Arthur loading for Texaco, 1914. From the Texaco Star, September 1914.

Above: A campsite near Spindletop.
COURTESY OF THE TEXAS ENERGY MUSEUM.

Below: Gladys City, the unplanned
Spindletop boomtown, in 1903.
COURTESY OF THE TEXAS ENERGY MUSEUM.

discoveries peaked quickly and declined sharply in less than two years from initial discovery, the string of finds kept Texas crude oil flowing on the local market encouraging outside investors to develop sizeable refineries in the upper Gulf Coast. It was the beginning of a new industry; one that continued to grow as the oil finds continued.

Refinery work was demanding and dangerous, but it paid well—much better than work on cotton farms or in gins and there was more of it. After Amoco developed a major improvement in refinery processes, the economics of that part of the oil business improved significantly by carrying the refiners through the downturns that followed World War I and changing the nature of work in their installations.

The strong demand for supplies and services encouraged such oilmen as Howard Hughes and James S. Abercrombie to build important new companies. Hughes Tool produced increasingly durable and versatile

drill bits while Cameron Ironworks met the demand for new valve and pressure control systems. Both companies improved their products steadily, and remained major employers in the Houston area—even after exploration shifted elsewhere in Texas.

From Spindletop to Sour Lake, the successive booms also familiarized Texans with social patterns that usually followed the discovery of a significant field. Lease brokers rushed in to sign up local landowners. Oilmen would buy these leases from them or from holdout farmers and ranchers. These oilmen would drill up the area as quickly as possible, and sell their properties while they were at peak level, and move on to new frontiers. In the meantime, drillers, roughnecks, tank men, unskilled roustabouts, service and supply salesmen and workers, and the men and women who housed and fed the oil field people surged into booming areas. They transformed crossroads villages and bare spots on the coastal prairie into bustling and noisy boomtowns, crowded and muddy, running around the clock every day of the week. Tours, (pronounced "towers") as shifts on rigs were called, ran twelve hours at a stretch, often with no days off until drilling declined in the field. In the meantime, the roughnecks earned three times more than they could have on farms and ranches. They also learned that it was hard to hang on to that money when they spent at least one-third of every paycheck to sleep in leaky tents, rough board buildings or crowded cot houses, and stand in line at oil field cafés for eggs, steak, and beans. They spent what little spare time they had in the bars and brothels that opened up almost as soon as the whiff of crude oil hit noses. Most of these workers were known by nicknames such as Heavy, Slim, Shorty, Red, and Blackie, obviously derived

from physical characteristics. Other names, such as Buttermilk, Teacake, Onion, Cabbage, and Corn were either taken from personal habits, as with "Buttermilk" and "Onion," or would often reflect worker humor, as with "Teacake," a notoriously sloppy eater. Some were usually good-natured ethnic jibes like an Irish worker being called "Cabbage." The origins of some nicknames, like that of Cesspool Willie, remain tantalizingly obscure.

During the early days in the Gulf Coast area, much of the dirt work and some of the laying of pipelines was done by African-Americans, but they were displaced by largely Irish crews early on. The same pattern held with other oil field work, as African Americans lost jobs as teamsters in the face of racial violence. In the Sour Lake Field, for example, Hughes & Davis hired out 150 teams all worked by black men until white workers threatened the firm with violence. Some new jobs were open to black Texans. In Beaumont and Port Arthur, African Americans were hired on at refineries, usually in warehouse and janitorial work, earning from

half to two-thirds as much as white men who tended stills and plumbed the plants.

There was also a permanently floating oil field population of drifters including unskilled workers, pipeliners, gamblers, rum runners, and prostitutes. As is still true in any settlement with large numbers of employed young men, the purveyors of thrills and pleasure did well, though they tended to engage in wars over territory and control, much like urban gangsters of the same era. Suicides, homicides, armed robbery, and most of the other serious crimes were committed by these elements of the floating oil field population even though journalists commonly blamed the "mud and gore" reputation of oil towns on workers. As exciting as the gory legends are, court records show that the young

Texans were more likely to drink too much and get into fights they couldn't finish. At that, the "urban legend" of oil town life, bodies in the gutter or in the river every day, is the created stuff of potboiler fiction—they were never common experiences of workers.

The greater threats to life and limb were on-the-job injuries, polluted drinking water, and spoiled food. Fire and explosion were ever present perils. When Curt Hamill, for example, went from Spindletop to drill at Batson, he saw an accident that was still on his mind a half century later:

> We lost a man…up in the derrick. The well blew out, the fire caught from a boiler possibly a thousand feet away and this man was up in the derrick. We didn't know whether he fell out or jumped out but he was burnt to death. And that wasn't easy to take, either. That's one of the saddest scenes I've had…seeing him lay out their—burn, boil in the oil when we couldn't get to him at all.

Hydrogen sulfide was a constant menace. Workers who did not detect it in time lost consciousness and died on rig floors if coworkers failed to rescue them in time. Even those who survived a face full of gas were left with swollen and burning eyes, which they treated with potato scrapings. William S. Farish, during his lifetime head of Humble Oil and Refining and of the Standard Oil Company of New Jersey, was overcome on a rig

Above: Out-of-town visitors flocked to the new oil fields.
COURTESY OF THE TEXAS ENERGY MUSEUM.

Below: A young society lady poses with black gold.
COURTESY OF THE HOUSTON METROPOLITAN RESEARCH CENTER.

floor, saved by a worker who pushed on his chest to expel the deadly gas. (In later years, Humble trained field workers in artificial respiration!)

And there was Beaumont water! Pure water was so scarce at Spindletop that it sold for $1 a gallon, more than twelve times the price of crude oil. Food was often inadequately refrigerated and kitchens and dining rooms often swarmed with flies. In one instance Howard Hughes and Jim Sharp were in a greasy-spoon restaurant, surrounded by swarms of black flies. Trying to turn the situation to a good end, Hughes, an enthusiastic gambler, suggested that they each butter a slice of bread, and the man with the most flies on his slice in one minute would be the winner. Unfortunately, there's no record of the outcome. It's not surprising that oil field doctors always identified diarrhea as the most common ailment. Workers usually referred to diarrhea as "the Beaumonts."

Despite dangerous working and living conditions, Texans survived to acquire experience as workers, supervisors, managers, and operators. In a way that was not characteristic of other states, the industry within Texas would thereafter have a distinctive Texan tinge, glamorizing the wildcatter and roughneck much as Texans had glorified the cattleman and cowboy of an earlier age as tough, resilient, and self-reliant. In folklore fiction, and history, they were all larger than life, a compliment denied to struggling farmers.

The developments that began at Beaumont also transformed the American petroleum industry, launching important new companies. These include Gulf, Texaco, and Humble (now Exxon USA). A decade before the United States Supreme Court broke up John D. Rockefeller's one-time monopoly, Jersey Standard, the flood of Texas crude sustained aggressive new competitors, initially Gulf and Texaco. Thereafter, Shell entered American markets, first in California then massively in Texas, Sun Oil grew, and Mobil was created from former Standard holdings by entrepreneurs from Galveston. When some of them moved offices from Beaumont to Houston, where Humble had its headquarters, they laid the foundation for Houston's status as the major management center for the American petroleum industry.

All of these monumental events began on that barren hill hear Beaumont on a cold day in 1901. Thereafter, exploration spread across so much of Texas that only two dozen or so of the 254 counties of its did not enter the list of producing regions. Texas would never be the same.

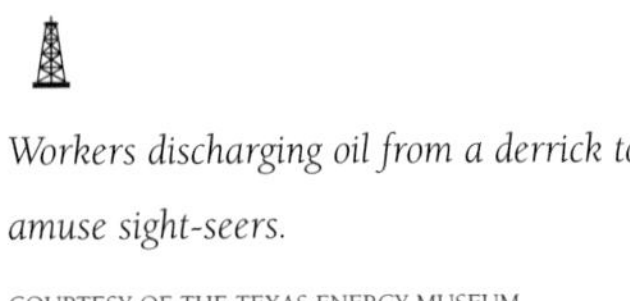

Workers discharging oil from a derrick to amuse sight-seers.

CHAPTER II

THE SEARCH SPREADS

"Damn the oil—I want water!"
- W. T. Waggoner

Beginning in the second decade of the twentieth century, wildcatters spread across Texas and found oil in most sections of the state. Beginning in North Texas, an area larger than many states, they tapped additional reservoirs, adding to the total production of the state, supplying growing refineries with feedstocks, and creating more jobs in the oil fields and elsewhere. Though its resources were already dwarfed by those of the upper Gulf Coast and would become less significant when measured against those of the Permian Basin and East Texas, North Texas provided the durable human images we associate with the industry and with oil booms. In part, credit—or blame—rests with Boyce House, a journalist who arrived in the region after the days of grand discoveries were history. Through a succession of books, and as history advisor to *Boomtown*, a 1930s film, House retold the tall tales old-timers laid on him, largely of the mud and blood variety—distorting the social life of the boom towns. House, however, wasn't alone in breaking into print with sensationalist folklore. Journalists who wrote for the most widely read magazines, *The Saturday Evening Post* among them, also found that bigger than life stories about Texas wildcatters boosted circulation. Black gold in Texas made good copy.

Though these writers distorted the reality of day-to-day life in North Texas oil towns with stretches of the truth and vivid language, they got one thing right: the region was largely dominated by independent oilmen, not associated with large companies that did everything from produce oil to sell refined products at the pump. In fact, the larger companies, Humble and Texaco especially, tried to solve the geological riddles of the area, losing millions of dollars in the process. From the beginning and to the end of the era of exploration, it really was the fabled wildcatters who brought in the major fields from Wichita Falls to Ranger.

Why didn't money and science win the day? Because nature just didn't cooperate. The producing formations in North Texas, often described as "stray sands," were only rarely located at this time by the newly important geoscientists. As luck would have it, surface geology commonly provided misleading indications of what would be found hundreds and thousands of feet beneath the surface of the earth. More than serious science, quasi-scientific "trendology" provided the incentives that led the smaller operators to drill. Simply put, trendology often involved nothing more than a surface map and a ruler. With the former, one could locate production or seeps, then pick up the ruler, and run lines from one to the other much like the child's game connect-the-dots. Most of the wells drilled on the basis of this highly flawed method were dry holes, but when oil

Wichita Falls, a new city in 1885.
COURTESY OF THE LESTER JONES COLLECTION, THE MUSEUM OF THE GREAT PLAINS, LAWTON, OKLAHOMA.

prices increased and wells could be drilled cheaply, incentives grew and risk shrank, encouraging small wildcatters to test their notions with the drill bit.

Even with a map and a ruler, it was not easy to locate a site for random exploration and the production in the region began with purely accidental finds—the result of random drilling. As was common in many parts of Texas, farmers and ranchers had found both surface seeps of oil and gas, and small amounts of oil in shallow water reservoirs. In the late nineteenth century, the owner of a livery stable in Wise County drilled a water well on his property only to find that the water was covered with an oil slick, though there was "less than usual," according to the local newspaper. One rancher, W. T. "Ole Tom" Waggoner, had a million-acre spread and a large herd in three counties. What he lacked was a reliable supply of water, so he drilled for it in Wichita County, and got more polluted water. He wasn't happy: "I said damn the oil, I want water." His perspective changed somewhat as years passed and income from leasing and royalties made him one of the wealthiest men in Texas.

Though seeps and natural pollution of shallow aquifers were never reliable indications of an underground oil reservoir or even enough oil or gas to produce, the numerous traces of oil encouraged some businessmen to believe that where there was a little oil, there might just be a whole lot more. Beginning in Electra and moving along to Burkburnett, Iowa Park, and other sites, a series of discoveries brought North Texas into the oil age. The entry was modest enough, especially compared to the Cushing and Healdton Fields in Oklahoma, but it was enough to sustain exploration, until another accidental discovery blew the lid off. This time, the Texas Pacific Coal Company, which drilled in search of coal seams in Palo Pinto County, brought in the Strawn Field in 1915. In response, civic leaders at Ranger to the west, offered the company a block of 25,000 acres in exchange for four wildcat tests. The company took the bait and, in October 1917, brought in a sixteen-hundred-barrels-of-oil-per-day (bopd) well just a mile out of town. The boom was on, moved along by a succession of wells that were exceptionally prolific for North Texas. For example, two years after its initial Ranger discovery, the Texas-Pacific completed a well that produced 11,500 barrels a day, discovering that the Ranger area had multiple producing layers or horizons. That knowledge led others to drill more wells, and to reenter weak wells and dry holes to seek the new reservoirs.

During the same decade, wildcatters drilled inside the Burkburnett settlement and discovered the Townsite Field, easily the most photographed in Texas history. The sight was startling, even compared to earlier views of Corsicana, because rigs appeared in back

Below: Wichita Falls as a mature cattle-cotton town in 1906.

COURTESY OF THE LESTER JONES COLLECTION, THE MUSEUM OF THE GREAT PLAINS.

Bottom: Wichita Falls during its oil age, 1919.

COURTESY OF J.C. AND W. F. REYNOLDS, WICHITA FALLS, TEXAS.

yards, on front doorsteps, all through town, several sited in a single small city block. By the time the field was half drilled, production from it and the other site in North Texas exceeded the capacity of pipelines and tank cars to remove it, so oil men there, as they had already done under similar circumstances in California and along the Texas Gulf Coast, dug vast earthen storage pits to hold the surplus crude.

As journalists reported once more, everybody and his dog wanted to get into the oil business in North Texas. In some instances, the dogs might have done better than some of the wildcatters. Hope counted for more than experience, never more clearly seen than in the history of the Lubbock Oil Company. Prominent local businessmen in that city, on the Plains, raised capital for the company then hired a local minister to manage the venture for them. As it would turn out, that hire was their first error in judgment. The minister then hired his brother-in-law, also new to oil, to supervise the drilling of a test well, a second mistake. Once under way, the new company missed the deadline for drilling contained in its lease, and lost $15,000 before their rig scratched the earth. Then, over a period of several years, the crew lost drilling tools down the hole, jammed pipe, and pulled the rig down when they tried to free it. They went on to burn up two tank car shipments of oil to fuel their boiler when they discovered that it refused to flow in freezing winter weather—a problem they undertook to solve by lighting a

fire under it. After more than a year in the business, the manager of Lubbock Oil saw shows of oil in a well—five miles away from their lease. Two years after their optimistic beginning, corporate officers liquidated the venture, a dead loss for them and for other investors. Similar tales of woe were enacted across the region by would-be oilmen who failed to operate and to limit their risks. By the time the North Texas fields came in, risk management usually involved drilling with "Other Peoples' Money" raised by bringing other wildcatters and investors outside the industry to share the risk of drilling a dry hole. Thus, though science did little in North Texas to cushion an oilman against total loss, sharing it with partners

Above: The familiar forest of derricks, this time scattered through Desdemona in 1919.
COURTESY OF THE J. A. "BOSS" BONNER COLLECTION, THE PERMIAN BASIN PETROLEUM MUSEUM.

Below: Another familiar oil town sight: teamsters, mules, and mud in Ranger, Texas, in 1918.
COURTESY OF THE PERMIAN BASIN ARCHIVES, UT-PERMIAN BASIN.

made it possible for him to survive to drill another day, in another location.

While Lubbock Oil was losing its shirt in North Texas, the large integrated companies were finding new fields and building their reserves of producible oil back on the Gulf Coast. Texaco and Humble did exceptionally well in the new West Columbia Field, part of

which underlay a plantation bought by former Governor James Hogg with oil money he made at Spindletop. Some of West Columbia's wells eclipsed earlier finds; Texaco's No. 49 on the Hogg estate, for example, blew in with twenty-eight thousand barrels per day. Even after wells in the field declined from initial high flows, something that usually happened well within a year of initial production, Humble was still producing twenty thousand barrels per day in the field and Texaco about half that amount.

Other discoveries in the region sustained production even after older fields declined. The large Orange Field, for example, kept Gulf, Texaco, and Mobil refineries well supplied. After evaluating often scant geological data, oilmen reentered older fields, including Goose Creek and Humble, drilling deeper, testing the flanks of the fields, and cleaning out old wells. At West Columbia, for example, Texaco deepened a 50-bopd well and boosted production from it to 2,900 barrels per day. In an extension of the Goose Creek Field, Simms-Sinclair completed a 35,000-bopd well and Gulf found a new pay sand with a 12,000-bopd well. Even Spindletop, often retested, without much success, received a new lease on life when Miles Frank Yount, T. P. Lee, and William Lee drilled a 1,500-barrel well followed by others that produced as much as 10,000 bopd. Following the familiar strategy of wildcatters, they held extensive leases when they drilled then sold their properties before production slumped. Yount-Lee sold their holdings to Amoco for more than $41 million. The rebirth of the historic field yielded even more black gold than the initial discoveries with Spindletop providing more than one-third of the production for the whole Gulf Coast region in 1927. (Thereafter the old field was reentered again, in 1951, and additional oil was found in yet another new horizon.)

During the 1920s, the result of the reentries and workovers, added to new discoveries at Barber's Hill, Pierce Junction, and Dayton, produced another tidal wave of oil, enough that the Texans who founded the Humble Oil Company decided to invest in what became a massive refinery at Baytown, near Houston, further expanding the Beaumont-Houston area as the manufacturing center for the Texas oil industry. The Baytown works were expanded greatly in future years as it became one of the largest refineries in the world. At Port Arthur and Beaumont, refiners also added to the capacities of their plants regularly.

By 1929, the Mobil and other refiners were processing more than three hundred thousand barrels of crude every day. The Humble refinery, opened at Baytown in 1919, was the largest in the country with a capacity of 140,000 barrels per day by 1940. During the lifetimes of the Hamill brothers, Humble and other refiners in the Beaumont-Port Arthur-Houston area employed almost one-third of all of the workers who worked in manufacturing.

Central Texas also boomed during the 1920s. Within seventy miles of the original Corsicana discovery, wildcatters brought in new fields that were far more prolific than the original find. A. E. Humphreys, one of the most colorful, worked with W. A. Reiter and Julius Fohs, both successful geologists, to reexamine scientific data that had led large companies to write-off the region. On the strength of their advice, Humphreys assembled a six-thousand-acre lease near Mexia, and drilled. His geologists were right! Humphreys' discovery well roared in on November 19, 1920, with a modest two hundred bopd; not enough to turn heads and change opinions, but sufficient to send Humpreys on a wild leasing spree as he tied up twelve thousand additional acres on the Mexia fault. As days passed and Humphreys drilled more wells, it was clear that he, Reiter, and Fohs were onto something big. Their second well produced four thousand bopd. A few weeks later, they struck it even bigger with Adamson No. 1, which had the potential for twenty-four thousand bopd. The boom was on. Leases near the first well, acquired for $5 per acre shot up to as much as $1,000. Once Humphreys built production on his leases to more than 40,000 bopd in the Mexia Field, the wildcatter sold out to the Pure Oil Company for $30 million. The play went on with Humble paying $2,000 per acre for a lease another independent had bought for $25 an acre.

Humphreys didn't stop to count his profits. He and Fohs spread out in the region to identify other hot prospects, settling on a location just ten miles from their great discovery, doing just as well with a ten-thousand-bopd find near Currie. Following the usual game plan of independents, Humpreys developed the close-in acreage and sold his leases to large companies before production began to decline. "Get in and get out" worked in Currie and Mexia as it did elsewhere. His next stop was near Powell, the site of a number of dry holes. This time, the pickings were less generous for Humphreys; though he held five hundred acres near the site of his discovery well, Humpreys missed making another killing because the larger companies, now more respectful of his geologists, had also blocked up sizeable leases. Gulf did especially well, with two wells on a single lease producing twenty-four thousand bopd.

To the west, near Luling, another wildcatter, far less qualified than Humphreys, Reiter, and Fohs, scooped up leases and drilled six dry holes before he struck oil. When he did, Edgar B. Davis, a shoe manufacturer from Brockton, Massachusetts, and his new company, the United North and South Oil Company, hit the front pages of Texas newspapers. As had been the case with Humphreys and Fohs, Davis rejected orthodox geological findings, and leased extensively in an area 7.5 miles long and more than half a mile wide. Like the Mexia wildcatters, he confounded the experts. By the end of 1924, the Luling Field was producing 47,000 bopd from almost 400 wells. Davis sold

Above: Storing crude oil at Humble during the 1920s.
COURTESY OF THE HOUSTON METROPOLITAN RESEARCH CENTER.

Below: Putting the lid on Mexia! Rangers and other lawmen pose for the camera after a big raid.
COURTESY OF THE TEXAS COLLECTION, BAYLOR UNIVERSITY LIBRARY, WACO.

some of his properties, keeping the rest to endow a foundation that has long supported charities in the region of his discovery.

Companies and wildcatters kept up the pace of exploration in other parts of Texas drilling along the Pecos River in far western Texas, along Corpus Christi Bay, and along with the shared border with the Republic of Mexico. The results didn't match those in North Texas and in the upper Gulf Coast. In Southwest Texas, cattlemen reported finding small amounts of oil, from the 1880s onward, and one of them, Robert Driscoll, was so enterprising as to drill all of the wells on his leases taking on the risks and reaping the profit. After his death, son Robert and daughter Clara continued to manage both the cattle and oil ends of the ranch business successfully, making them rarities in Southwest Texas.

Near San Antonio, drillers located a small field near the missions, but the yield was small, too modest to compete for the exploration capital that went into other regions. During the 1920s, however, explorers found the Somerset Field, some distance from town, just large enough to support one large local operator, Grayburg Oil, headed by F. L. Thompson, a medical doctor. The shallow wells at Somerset were cheap and quick to drill, taking less time than a patient did to recover from an appendectomy.

Near Corpus Christi, the search for oil led to the discovery of several large natural gas fields, more than big enough to supply the city and towns in the region, but not sufficient to start a boom. At pennies per thousand cubic feet, gas was a loser's game and it was produced only when it was discovered during searches for crude oil. Still, some of the gas wells provided local drama, as when one blew in so strongly that it forced gas out of neighboring dry holes and produced gas geysers in Corpus Christi Bay.

Though the region would later yield vast amounts of gas and crude oil, through the 1920s there were few large discoveries. More typical were the finds of O. W. Killam, a colorful and inventive wildcatter in the Laredo area. Killam, who had promoted sales of mining stock before he came to Texas, signed leases for little or no money with local landowners who were eager to have some of those fabled gushers on their land and to receive "royalties"—usually payment for one-eighth of the oil produced under their fields. As promoters went, Killam was both honest and successful, risking his own capital, managing carefully, and realizing more than modest success. Over a ten-year period, he

discovered nearly a dozen small fields in the region following hunches based on surface geology and the lure of cheap leases. He drilled relatively few dry holes, a fact that became something of an embarrassment because he promised his driller, Joe Morris, a new hat for every successful well. Early in their relationship he called off the bet, on the grounds that Laredo shops had run out of hats in Morris' size—and that the hats might be worth more than the wells.

Though the total production from Killam's fields was small by Texas standards, it was big enough to take him and his investors into the pipeline and refining ends of the business. Even more important, his sustained success kept interest in the region alive. Over two decades, he put his corner of Southwest Texas on the oil map, but he also seemed to prove that the gushers would have to be found elsewhere.

One more area of Texas, the vast Panhandle, came into oil play during the second and third decades of the twentieth century. Amarillo businessmen and several prominent ranchers decided not to wait for the large companies and North Texas wildcatters to find their black gold, so they formed the Amarillo Oil Company and began to drill on the basis of advice from a prominent Oklahoma geologist. Their first find, on the Masterson ranch, was

the discovery well in what proved to be a massive natural gas field, underlying nearly five million acres. Without a sizeable local market, payout on Amarillo Oil's investment awaited completion of natural pipelines in the next decade. The big returns were in the future. By 1994, there were about forty-five hundred producing wells in the Panhandle Field, the largest volume natural gas field in the United States. In the meantime, the magnitude of the first gas well and those that followed it convinced oil men that where there was that much natural gas, there had to be lots of oil.

Experience elsewhere often swore at that belief, but in the Panhandle, it prompted the sustained exploration that led to the opening of

Above: Panhandle cowboys portray old-time Texas as they ride by a steel derrick on the SMS Ranch.
COURTESY OF THE C. C. RISTER COLLECTION, SOUTHWEST COLLECTION, TEXAS TECH UNIVERSITY.

Below A new-time Panhandle site, the Phillips refinery and camp at Borger.
COURTESY OF THE DR. P. J. EDMONSON COLLECTION, PERMIAN BASIN PETROLEUM MUSEUM.

the Borger Field and to smaller fields in Hutchinson, Carson, and other counties. As the oil play unfolded, it became clear that oil-yielding geological formations were deep, beyond the resources of most independent oil men. That left the first round of development to larger firms, including Phillips and Skelly, both of which also built extensive company camps to house their production and processing employees. Once Humble, Gulf, and Mobil built oil pipelines into the region, crude prices rose, prompting further exploration.

Borger's boom days were much like those of Ranger, Mexia, and other oil field towns during the 1920s. Ace Borger, a real estate promoter from Chicago, bought the townsite and sold lots to the highest bidders without much concern for such conventional comforts as sanitary plumbing, clean water, and tidy streets. The new town, which Ace named after himself, grew quickly as leaseholders rushed to drill wells. The influx of workers and service personnel was generously expanded by the arrival of professional criminals from Currie and Mexia, where Texas Rangers and federal revenue agents had made life hot for them with repeated raids and round-ups. The Rangers showed up in Borger for a repeat performance staying just long enough to stage dramatic arrests, pulling in gamblers and moonshiners along with roughnecks who drank too much and fought each other. The latter were usually released without formal charges after they had spent twenty-four hours chained to hitching posts, but some of the professional criminals were held for arraignment. That done, the Rangers cleared town, leaving local lawmen to keep a lid on the most dangerous violations of the law. They succeeded, for the most part.

Though Borger boomed in the way of new oil towns, the permanent gains went to the regional management center, Amarillo, where the influx of newcomers more than doubled the city's population during the 1920s. Thereafter, Amarillo sustained a small but vital oil and gas community through successive booms and busts.

The second and third decades of the twentieth century were periods of dynamic growth in both pipeline construction and processing across Texas. One new field followed another, and the large companies built new refineries in North Texas and in the Panhandle. Less visible, but of even more importance to the future of the state, they continued to expand and improve their installations on the Texas Gulf Coast to the point that the largest concentration of refinery works in the United States was within one hundred miles of Houston. As they built, Beaumont, Port Arthur, and Houston continued to grow. By the middle of the 1920s, wildcatters had found oil in most parts of Texas, but not in Permian Basin and plains areas of West Texas and not in East Texas, all of them apparently unpromising territory for the drill bit. Time would tell.

Tank farms, pipe yards, and derricks in Borger, 1926.

CHAPTER III

OIL ON THE DUSTY PLAINS

"Unforgiving skies, baking heat, and rainlessness"
- Larry McMurtry

Most parts of Texas were explored by the middle of the 1920s without impressive results in the Permian Basin, a vast semi-arid expanse in far west Texas, a region Larry McMurtry aptly described as one of "unforgiving skies, baking heat, and rainlessness." It was the next stop for the intrepid explorers, even as new wells gushed elsewhere. There was little to encourage speculative drilling in that vast area at the beginning of the decade. The current stage of geology, involving correlation of surface to subsurface formations and analysis of minerals brought to the surface during drilling, were of little use because a good bit of the area was covered by shifting sand. So little drilling had been done in the Basin that there was no collection of drilling logs and cuttings. For that reason, large companies and independents alike avoided the area. Their reluctance was reinforced by the scant history of drilling, which had yielded finds of no greater commercial importance than those at Nacogdoches sixty years earlier. During the 1920s, that changed as the result of three "rank wildcat" discoveries, far from producing oil fields.

In the far eastern part of the basin, near Colorado City, one wildcatter brought in the Westbrook Field at twenty-five hundred feet in 1921. It was the first field in the vast territory, but the new discovery did not lead to additional activity because oilmen generally wrote it off as an anomaly, not reading the future

The rig and tanks at Roy Westbrook's discovery well in 1928.
COURTESY OF THE PERMIAN BASIN ARCHIVES, UT-PERMIAN BASIN.

Top, left: Rupert Ricker, a Permian Basin oil pioneer.

Top, right: Frank T. Pickrell, one of the organizers of the Texon Oil and Land Company, 1925. This photo originally appeared in the San Angelo Standard-Times.

Right: Haymon Krupp, the money behind the discovery. This illustration was made by Jose Cisneros.

from that isolated find. Orthodox opinion still thought it unlikely that there was much oil west of Ranger, on the vast plain that stretched for more than three hundred miles.

It took more imagination than science to view West Texas as a area in which fortunes might be made in the oil business. The idea of making money from the prospect of oil under the ground came to Rupert Ricker, a bored young lawyer who hanged out his shingle in Big Lake, a small market town in the southeastern part of the Permian Basin, then waited for his neighbors to sue each other. With lots of time on his hands, Ricker imagined a way to make a great deal of money without running the risk of drilling dry holes. He would lease University lands for a dime an acre, then sell leases for $1 to $5 per acre to anyone willing to take a chance. When that dream became a plan, Ricker acquired a lease on 675 square miles for $43,136. If he sold leases at even a dollar an acre, he stood to make a small fortune.

Once he was committed to the idea, Ricker still needed ready cash to pay the state for his leases. After he failed to find local backers, Ricker turned to Frank Pickrell, an Army buddy who had some experience in the oil industry. Together, they traveled to the East coast and through the Middle West in futile attempts to unload their abundant acreage. The pair found that there was no longer

strong demand for either leases or for shares in oil companies because the Fort Worth trials of such fraudulent promoters as Dr. Frederick A. Cook and Seymour Ernest Jacobson Cox, also known as "Alphabet" Cox, well publicized in Eastern newspapers, had drained the pool of investors in visionary oil schemes. Though the two Texans were not scam artists, experienced investors were wary because the real gushers were producing in the Gulf Coast area, in north Texas, and in

central Texas. There was hardly any need for oilmen to venture into the arid West Texas area where the nominal river, the heavily alkaline Pecos, had long been known as "the graveyard of cattlemen's hopes."

As a last resort, Pickrell contacted Haymon Krupp, a prosperous El Paso merchant. Krupp was experienced in speculative mineral investments, including the Santa Rita copper mine in nearby New Mexico. From his point of view, the venture was a long shot, but it did include an immense amount of acreage and that asset alone made the deal attractive, so Krupp bailed the two speculators out and paid the lease costs.

With his backing, a trial well was undertaken, sited as close to the railroad as possible to speed up construction and drilling. As luck would have it, on their first try, Santa Rita No. 1, came in as a one-hundred-barrel well on May 28, 1923. At the time, a well that size excited little immediate interest within Texas oil circles, and Ricker, Pickrell, and Krupp lacked sufficient experience and capital to develop the vast leasehold so they recruited Mike Benedum, a veteran West Virginia wildcatter, to take over the venture and drill additional wells. Even with Benedum at the helm, the venture attracted little favorable notice until it was well along. Then, Santa Rita No. 9 and No. 11 came in as bona fide gushers producing 1,500 and 3,000 barrels per day, just when central Texas wells were tapering off. By 1925, the Big Lake Field was on the *Oil and Gas Journal*'s map, producing more oil than tank cars could accommodate economically, so Benedum signed a purchase contract with Marland Oil, a medium-sized integrated firm. With that agreement, Ricker's dream, Pickrell's effort, Krupp's investment, and Benedum's

Above: Robert R. Penn, discoverer of the Pennwell Field, near Odessa. Highly active in oil industry associations, he is posing for this photograph at the American Petroleum Institute golf tournament at its Dallas meeting in 1929.

Left: Teamsters, mules, and pipe on the endless plains of the Permian Basin.

management paid off, opening the vast Permian Basin to exploration for black gold.

A third random discovery sustained interest in the region. In that instance, Chester Bunker, a Fort Worth printer, gave away interests in a lease he acquired on the Lewis Powell ranch, less than twenty miles from the Big Lake Field, to new subscribers to his promoters' tip sheet, "World Oil"—a name that a later and unrelated trade journal assumed. Bunker's gimmick worked, but his timing was bad because, just then, post office inspectors were scrutinizing Fort Worth promoters, judging whether they kept their promises, and bringing suit against some of those who had not. With scant choice, Bunker drilled a well and brought in what he called the World Field. Accident or not, the oil was real, the third find in the area, which attracted the attention of better-heeled independents and of large companies.

Humble, Gulf, and other companies invested in geophysical studies of the vast region, their paleontologists and geologists studying cuttings, cores, and drilling logs from completed wells. In response to optimistic reports from their scientists, the large companies launched extensive leasing campaigns, tying up vast acreage easily because large ranches dominated the arid area. By the end of 1924, leasing had moved to adjoining counties. To the north of the Big Lake and World Fields, Marland gave George B. McCamey, a Fort Worth drilling contractor, acreage to pay for a test well. Once he obtained additional leases from Amoco, McCamey sold enough of them to third parties to pay his costs for the test. Another risk paid off when he struck oil at 2,165 feet, enabling McCamey and his partner, J. P. Johnson, to sell their leases for $500,000 and move on to new prospects.

To the west of McCamey's discovery, two other independents, Robert Fields and George M. Church, discovered another sizeable field in Crane County in 1926. They also sold out, in this instance to Mobil, for $2 million.

Above: Wink, the new boomtown under construction in the 1920s.
COURTESY OF THE PERMIAN BASIN ARCHIVES, UT-PERMIAN BASIN.

Below: Teamsters moving bunkhouses to Wink in 1928. The photographer touched up the lower part of the photo to add vegetation to the barren environment.
COURTESY OF THE NOLAN COLLECTION, PERMIAN BASIN ARCHIVES, UT-PERMIAN BASIN.

Five miles away, on the McElroy ranch in Crane County, Gulf bought extensive leases from the McElroy Ranch Company, which had purchased 79 sections of land and 6,000 head of cattle from the original rancher for $2.5 million. It turned out to be a good deal for Gulf. During 1926, the company completed its McElroy No. 1, the first well it drilled on its McElroy leases, for a steady six hundred bopd. Additional drilling produced so much oil that Gulf invested in four hundred thousand barrels of storage in the field and continued to develop it well beyond the end of the decade.

To the north of the Big Lake Field, wildcatters found still more oil in Howard County, beginning in 1926 with the Chalk Field and continuing through the opening of the large Dora Roberts Field. The latter stirred up considerable excitement in Big Spring, the nearest town, when Magnolia drilled deeper in one of its modest producers and got twenty-four hundred bopd. Enough to fuel a local boom and make more ranchers rich from lease bonuses, rental payments, and royalty income. Dora Roberts, for her share, left an $8-million estate when she died in 1954.

The biggest Basin finds of the decade came during 1926, with the discovery of the Yates and Hendrick Fields. Yates was the first giant field, one that would pass the one billion barrel mark, with hundreds of millions to go. The story of the Yates Field illustrates the importance of both science and financial staying-power. Transcontinental Oil, run by Mike Benedum, took out leases in Pecos County on the fifty-thousand-acre Yates ranch following the discovery at Big Lake. Then Transcontinental and Mid-Kansas, an Ohio Oil Company subsidiary, set their geologists to work, seeking a scientific basis for risking money on the leases. Accepting their optimistic report was chancey at that time because even university-educated geologists were often regarded as optimists, more attuned to the pluses than minuses, when they came to evaluate data. The joke was often told, in which one geologist met another in an oil town. The first scientist trumpeted "the well was a great success." His knowing friend added, "But you didn't find oil!" What the first geologist was telling the second was that a significant part of his interpretation of data was vindicated and that further exploration might lead to discoveries in the future.

Knowing that even the best geologists could be wrong, the two firms cooperated to drill

Wink in 1929.

Above: Drugstores, cafés, and grocery stores on the Main Street of McCamey, near the Yates Field.

COURTESY OF THE ABELL-HANGER COLLECTION, PERMIAN BASIN PETROLEUM MUSEUM.

Below: The new boomtown of Crane, near the Church and Fields discovery, south of Odessa.

COURTESY OF THE PERMIAN BASIN ARCHIVES, UT-PERMIAN BASIN.

four test wells on the basis of the best guesses of their scientists. The project didn't begin well: wells one and two were dry holes. Moving to the Yates dome, two miles from the salty Pecos River, the companies drilled to about 1,000 feet and got a 2,200 bopd gusher in late October 1926. Chevron drilled on nearby acreage, completing a 2,200-barrel horizon, but its men kept on drilling until they tapped a deeper reservoir for 122,000 bopd. Gulf kept the fever going with a well that was capable of producing 18,000 barrels, then lost pride of place to Sims Oil, an independent, which had one well that flowed 86,000 bopd. The giant of the field, No. 30A, came in for more than two hundred thousand bopd, leaving its drilling crew soaked in crude, stunned, and happy. Early on, it was clear to the dozen or so

operators with property in Yates, that it was truly a bonanza field. Before the 1920s ended, oilmen were producing from more than twenty thousand acres of it. By then, producers in the Yates Field had agreed to a voluntary limitation of their production, in the interest of long-term production and profits.

The significant if less spectacular Hendrick Field discovery was drilled by a Fort Worth promoter Roy A. Westbrook who was under the eye of postal inspectors. Having carved up acreage on the Hendrick ranch into 850 five-acre tracts, Westbook had sold them off on the basis of his promise to drill a test well. Like Chester Bunker one year earlier, federal agents watched his moves. Westbrook had to make good on the pledge and he did, with a discovery well that flowed 150 bopd. As interest in the Basin mounted, even a relatively modest well was enough to set off another lease rush; Gulf, Atlantic Refining, Shell, and other companies hurried to get in on the new Hendrick Field. By the end of 1926, there were bigger wells—Gulf's first test was good for 2,000 bopd—with later wells coming in at more than 25,000 bopd. Thereafter, the news continued to improve when Shell completed a 60,000-barrel well, and Gulf and Atlantic both put 50,000-barrel wells on line. The rancher who signed the leases sold his holdings for $3.368 million, giving away about half of that amount to charities and investing the balance in urban real estate in Corpus Christi and Abilene.

Like so many other fields, the Hendrick was extended, notably in 1928, when Sid

Richardson, a Fort Worth operator and driller, found production six miles south of the discovery well. The next year, Richardson drilled on a farm-out from Shell 22 miles from the Hendrick Field and got a 1,000-bopd well on the Scarborough ranch. By the beginning of 1929, there were 582 wells in the Hendrick Field operated by 32 producers. The biggest, Southern Crude, held 193 of them. The potential production for the new field was 660,000 barrels of oil per day, but the limitation of pipeline capacity reined it back to 150,000 barrels. Like the Yates Field, Hendrick's production was restricted, but negotiation of a formula (lease-holding and production per well) took longer mainly because there were more producers in the Hendrick. The independents were rarely able to sacrifice current income for future profits because each one operated on a shoestring, and the string was usually badly frayed. As production from wells in the field declined, a voluntary restriction program was signed to be enforced by the Texas Railroad Commission.

Despite the sharp rises and falls in the price of crude oil, as the 1920s came to an end, there were more than a dozen significant oil fields in the Permian Basin. Large companies had acquired both extensive leases inventories and a growing mountain of geological data that could be used to unlock the secrets of the vast stretched of undrilled territory. Humble held more than 1.75 million acres in the region, followed by Gulf with 828,000, Marland with 738,000, and Mobil with 452,000. Shell, Pure, Amoco, Sun, Chevron, and Sinclair also held strong lease positions. Tremendous sums had been invested in the Basin by 1936: $90 million to drill nearly 2,000 producing wells and 1,300 dry holes; storage tanks had cost more than $25 million, pipelines at least $75 million and annual rents paid to landowners passed $10 million. And that was when lunch at an oil field café cost twenty-five cents! The vast region had about six thousand miles of gathering lines within oil fields and pipelines between fields and termini. With cheap and abundant crude to be had, Humble, Col-Tex, and other companies built refineries to process some of that crude oil. In just six years, the Permian Basin emerged as one of the most important producing regions in the United States. In future decades it would become the principal oil province in the lower forty-eight states.

In the meantime, the sparsely settled area experienced its first oil booms, beginning

Above: Workers were lucky if they could get rooms at the new 12 Oaks Motel in Odessa in 1927.

Below: The road to Crane from Odessa, the new oil field service center of the Permian Basin, in 1930.

Top: Patrons even slept on cots in the halls
of Odessa's new Elliott Hotel in 1929.
COURTESY OF THE PERMIAN BASIN PETROLEUM MUSEUM.

Above: The upscale lobby and cigar stand
at the Scharbauer. The hotel's coffee shop
was long a popular site for oilmen's
business breakfasts.
COURTESY OF THE PERMIAN BASIN PETROLEUM MUSEUM.

lease payments, rents, and royalty settlements. With caution born of earlier adversity, he added a few luxuries, notably electricity and new appliances, but he was in no rush to change his life style. So, it was with some misgivings that he abandoned long underwear for sleeping, grudgingly donning a night shirt, brought from Dallas by his wife. He never quite appreciated this part of the upgrading oil money brought into his life; the innovation did not last long.

On one memorable night, as the story has it, Pink was sleeping soundly when his wife shook him: "Pink, wake up! There's a ruckus out back and I think critters have gotten into the hen house again."

The weary rancher got to his feet slowly and plodded through the house, Mrs. Mitchell right behind him. At the back door, Pink grabbed his shotgun—his dog joined the procession, right behind Pink. Once the parade of rancher, dog, and wife reached the hen house, Pink threw the door open. Before his eyes could adjust to the dark, he let off a blast from his gun, sending hens flying in all directions. Mrs. Mitchell, startled by the noise and flying feathers, nudged him for an explanation. "Pink, why in the world did you do that? You've killed off at least half a dozen of my best layers."

Stoically, Pink responded: "My dear, you'll never know how cold a dog's nose can be...." Pink went back into the house, threw away the new night shirt, and slipped into a pair of cotton longies.

Gentrification of the old timers came slowly, often grudgingly. The new oil money, however, was always welcomed.

The bulk of the boomtime population was made up of ordinary working people and their families. First on the scene were the leasing agents, who usually found rooms in small hotels. After them came drillers, roughnecks, and roustabouts, who did the often dangerous work of drilling and completing wells. The floating oil field supply and service people came right on their heels. Drilling and service workers and, sometimes their families, often lived in makeshift housing—commonly war surplus tents, with plywood or boards four feet up on each side

with Big Lake and moving westward steadily, month by month. In its wake, the boom transformed the region and changed lives. Ranchers, many of them landpoor, were the first to benefit. Some of them, like Ira Yates, sold out and moved to other parts of Texas. Others stayed put, paid off their debts and adjusted to relative affluence. Sometimes that process was slow. In Winkler County, for example, "Pink" Mitchell gradually amassed a small fortune from oil

and on the floor. Even as new towns such as Crane and Wink grew, these tent cities continued to expand on their outskirts and in the oil fields, providing cheap, primitive, and portable housing.

In the Permian Basin, tent dwellers were barely sheltered from the oppressive heat, constant winds, and blowing dirt and sand. Still, they made do trying to live conventional lives under exceptional circumstances. One lady moved half a dozen times in as many years, always wrapping her prized wedding gift—a crystal punch bowl—in soft clothing. It never served its original function, becoming a fish bowl to amuse children on the second move. Other wives also did their best to preserve what they knew as normal family life, cleaning, cooking, raising children as nearly like what they would have done on farms. The women baked and cooked with special effort for holidays. One woman used a two-burner kerosene stove to make a full Thanksgiving dinner: turkey, dressing, mashed potatoes, canned peas, Jell-o, and a pumpkin pie for dessert. While her family was enjoying the special treats, a West Texas dust storm closed in, forcing them to finish dinner under bed sheets, right down to gritty pie and sandy Jell-o.

These exploration workers and their families generally accepted the dirt because the most likely alternative to hard living in

Above: The Big Lake Oil Company's model company camp in 1928. It housed hundreds of workers in cheap and comfortable bungalows and bunk houses. The baseball diamond, upper left, and theater, center right, were highly popular.
COURTESY OF THE CHARLES BEYER COLLECTION, PERMIAN BASIN PETROLEUM MUSEUM.

Below: Big Lake Oil provided accordions and harmonicas for the school band.
COURTESY OF THE BILL THOMPSON COLLECTION, PERMIAN BASIN PETROLEUM MUSEUM.

Right: Shell's million-barrel tank in Monahans was built to hold abundant Permian Basin crude oil, but it leaked.
COURTESY OF THE PERMIAN BASIN PETROLEUM MUSEUM.

Below: The drilling crew on the Ohio Oil Company's Yates No. 30A just after the well came in at two hundred thousand barrels per day.
COURTESY OF THE PERMIAN BASIN ARCHIVES, UT-PERMIAN BASIN.

the oil fields was scratching a living on a tenant farm.

The more privileged production workers arrived after development of a field was underway. They were better housed than the more transient drilling personnel because companies knew that they would not retain valuable employees who had to cope with tent life. The solution, begun largely in the Gulf Coast area and in North Texas, was the company camp. These settlements varied in size and comfort, ranging from Big Lake Oil's model city to a dozen four-room houses lined up on the barren prairie. The Big Lake camp contained solid homes with at least two bedrooms and screened-in front porches. Once the buildings were completed, the company spent thousands of dollars on landscaping, including the creation of prize-winning rose beds. The Big Lake Oil Company also built schools, recreation halls, a large swimming pool, and a baseball diamond and stadium, along with bunk houses and dining halls for single workers. Like other large companies, Big Lake Oil also sponsored company baseball teams, one of them a national champion in amateur competition.

Typically, rents were nominal, deducted from paychecks, and utilities were gratis. The Humble camp at Cisco was a small city, with all of the usual amenities, including indoor plumbing. Small wonder that most of the people who lived in those camps remember them fondly. Over the next two decades, camps and tent cities dotted the landscape as oilmen located and developed the immense petroleum resources of the Permian Basin.

CHAPTER IV

BONANZA AND THE LAW IN EAST TEXAS

"Producing oil and gas sands in the fields now developed in the region surrounding the Joiner well [will] yield large gushers."
- A. D. "Doc" Lloyd, M. D., Ph.d., G.C.E. (a.k.a. Joseph Idelbert Durham), 1930

Drilling in the Permian Basin, along the upper Gulf Coast and in Southwest Texas slowed dramatically during 1929 in response to the opening of the Oklahoma City and Van Fields. The latter was developed relatively slowly because many of the leases closest to the discovery well belonged to the Pure Oil Corporation. With commanding acreage and declining prices, Pure was prepared to proceed slowly and await long-term profits. After Texaco discovered the Darst Creek Field, east of San Antonio, it was also carefully controlled because Humble, Gulf, Mobil, and Texaco owned most of the productive acreage and placed it under the Railroad Commission—supervised voluntary proration. The Oklahoma City Field, however, was wide open for most of the year, driving prices below $1 per barrel because of its accessibility to pipeline systems and major refineries. Then, as the nationwide depression cut into demand for refined products, prices of crude oil slid even farther. It seemed to many expert observers that the situation could not get much worse for oilmen.

"Dad" Joiner and "Doc" Lloyd shake hands at Daisy Bradford #3. H. L. Hunt, wearing a straw hat, looks on (center, ,right).

Above: Well-dressed men and women watch

a well completion in Kilgore.

COURTESY OF THE EAST TEXAS OIL MUSEUM, KILGORE
COLLEGE, KILGORE, TEXAS.

Below: Hauling a boiler in the East

Texas Field.

COURTESY OF THE PERMIAN BASIN ARCHIVES, UT-
PERMIAN BASIN.

yield oil very far from Corsicana in central Texas. Major producers, including Humble and Texaco had tested the sand during the 1920s without success. The Van Field, producing from the same formation, seemed to indicate that whatever production Woodbine contained would probably be found north and west of Tyler and Longview. As the depression deepened and money got tight, Gulf and other large leaseholders in the unpromising area let leases lapse, waiting for more promising prospects to drill when the economy improved.

Though both common sense and experience swore at the likelihood of significant reservoirs in the northeastern part of Texas, optimism and desperation won out. One of the remaining risk-takers, Columbus Marion Joiner, had enjoyed minor success in Oklahoma, but his principal asset was his ability to sell leases to naive investors, which he is supposed to have done by wining and dining prosperous widows, much like Max Bialystock, the character played by Zero Mostel in a 1960s film, *The Producers*. Apart from this exceptional approach to capital formation, Joiner's only asset was the advice of a would-be geologist "Doc" Lloyd. Lloyd claimed to have a full alphabet of academic degrees. He was, in fact, a pharmacist, entirely self-taught in geology and other fields in which he claimed expertise. Even without a college degree in geology, he could patch together imaginative pseudo-science, just convincing enough to win over investors who knew even less geology than he did. Whatever his limitations as a geoscientist, Lloyd was highly useful to Joiner. When the promoter brought Lloyd into his East Texas project,

They were wildly wrong. It is easy to understand why it seemed that oil prices had bottomed out by mid-1930. By then, production in the Oklahoma City Field was limited and there were no new bonanza fields on the horizon. There did not seem to be any reason for experienced oilmen to pay much attention to random drilling. O. W. Killam, among others, was still bringing in tolerable wells—in the one-hundred-barrel range—but that volume offered little incentive for wildcatters while crude oil prices were low. Moreover, some of the oilmen who persisted were seen by industry veterans as either flagrant promoters of ill-advised schemes or as gamblers, looking in highly unlikely places, such as Northeast Texas, where the Woodbine sand produced water abundantly, but failed to

"Doc" claimed that the leases overlay numerous anticlines, a fault line, and a salt dome, naming nearly all of the oil-trapping geological structures in North America. If anyone had ever found oil by drilling down through a kitchen sink, it is likely that he would have included the American Standard structure as well.

With Lloyd at his side, "Dad" Joiner began his search for oil in East Texas in mid-1927 on Daisy Bradford's farm. His first try failed because pipe jammed in the hole, a common mishap. The following year, he raised more money, obtained more supplies and labor on credit, and tried again, with a similar outcome. Joiner went on to a third well, taking more than a year to drill to thirty-five hundred feet by September 1930. When he reached target depth, the well, closely watched by both independents and major companies, spewed mud, gas, and a little oil, but not enough to bother with. However, the appearance of oil-soaked sand was enough to start a modest lease rush as H. L. Hunt and other independents hurried to acquire cheap leases, as close to the Joiner well as possible. The major companies picked up even cheaper leases, more distant from Joiner's well, expecting that the show of oil would be as much as the well would make. Many seasoned oilmen dismissed reports of oil shows from Ed Lasater, Joiner's driller, claiming that they were the results of salting, dumping crude oil

Above: The East Texas Field in 1932.
COURTESY OF THE PERMIAN BASIN ARCHIVES, UT-PERMIAN BASIN.

Below: Steel derricks in Kilgore.
COURTESY OF THE EAST TEXAS OIL MUSEUM, KILGORE COLLEGE.

trucked in from elsewhere to support Joiner's fundraising. Though he knew full well that Joiner and Lloyd had produced only hot air to date, Lasater's skepticism weakened as more oil-stained sand appeared. It was just possible, if not highly probably, that Daisey Bradford No. 3 would be a Texas gusher and that Joiner had a producing well. Lasater took time off to buy leases with his own money.

The improbable discovery came on October 3, when Daisy Bradford No. 3 blew in, spraying Joiner, Lasater, and the countryside with crude oil. Joiner claimed that the well was producing 5,900 barrels per day, a figure he boosted to 6,800 in a few days. The truth was

Right: The tangle of derricks, traffic, and
power lines on Kilgore Street in Kilgore.
COURTESY OF THE EAST TEXAS OIL MUSEUM,
KILGORE COLLEGE.

Below: The First Presbyterian Church
in Kilgore.
COURTESY OF THE EAST TEXAS OIL MUSEUM,
KILGORE COLLEGE.

NO. KILGORE ST. — KILGORE, TEX.
41.

more modest; the well settled down to flowing a little more than two hundred barrels a day by the time anyone more objective could assess it. At that, major companies were still suspicious. When Paul Davis, an oil scout for one of the biggest oil companies, telephoned his boss in Houston with news of Joiner's well, the supervisor asked Davis if he had inhaled too much gas! Other skeptics reasoned that Joiner's discovery would turn out to be a one-well field. The doubters did not hold center stage long. Two months later, a deep rock oil test a mile away produced 3,000 barrels; twelve miles beyond it, Ed Bateman got 22,000 barrels when His Lou Della Crim No. 1 blew in. Five weeks later, John E. Farrell, W. A. Moncrief, and Ed Showers completed Lathrop No. 1 for twenty thousand bopd. The boom was on.

Over a two-year period, a dozen distant wells defined the massive East Texas Field, which was 42 miles from north to south and from 5 to 12 miles wide, six times the extent of the Yates Field. The field spilled into five counties and produced abundant high gravity oil, the kind that yielded more gasoline at refineries.

The new field was drilled quickly. Following the legal principle of "capture," the oil belonged to whoever produced it, even if it flowed from under leases that adjoined a well. The smart thing for an oilman to do, obviously, was to drill as many wells as possible on his leases, before operators on adjoining leases produced crude from under them. From Spindletop onward, the law of capture had driven the frenzied rush to drill. Rapid drilling on small leases had always struck economists as wasteful, partly because it

was assumed that more wells depleted gas or water pressure rapidly and that less oil would be produced under the circumstances than would be with fewer wells. Like many common sense assumptions, this one would be greatly modified when petroleum engineers had more experience with aging wells, but the economic core of it was still solid: in any event, the same amount of oil could be produced with fewer wells, so the waste was economic.

Economics and science seemed grandly irrelevant as independents made the most of their opportunities, drilling wells cheaply, usually for about $25,000 each, then selling leases before production declined. Ed Lasater, for example, sold his well and two sections of leases to Humble for $2.1 million. He would never work for groceries again. As production climbed in the East Texas Field, H. L. Hunt

Above: Tank cars and pipelines carried the flood of East Texas oil.
COURTESY OF THE JACK NOLAN COLLECTION, SOUTHWEST COLLECTION, TEXAS TECH UNIVERSITY.

Below: A tent city at the Magnolia (Mobil) Camp near Kilgore, 1931.
COURTESY OF THE EAST TEXAS OIL MUSEUM, KILGORE COLLEGE.

bought "Dad" Joiner's leases, drilled wells, and built a gathering system in the giant field, carrying crude to tank cars loading racks for shipment to Houston and Arkansas. Hunt's pipeline could carry only a falling fraction of East Texas oil, so larger companies linked up their growing leases to their refineries on the Texas Gulf Coast.

By the end of 1933, oilmen had completed nearly twelve thousand wells in the field. Many of them were on small leases, some irregularly shaped like triangles and diamonds. Most of those leases contained as many wells as could be drilled on them. By that date, the field had produced more than 450 million barrels of oil, enough to drive prices as low as three cents per barrel on some bleak days. At that price, oil could not be produced at a profit anywhere; the barrel was worth more than the oil. Still, for oilmen who were drilling wells, there was no option to completing them, so the race went on. Three years later, there were more than twenty thousand wells in the East Texas Field.

The only part of the industry that could make money on dime oil was refining because the cost of crude fell much faster than the prices of gasoline and other products. Humble and other large companies stored as much cheap crude as they could, but limited storage and pipeline capacity left a lot of oil available, for the taking in the oil field. This opportunity led small businessmen, some of them oil producers, to build plants, some no larger and no more advanced in technology than moonshiners' stills. By September 1931, there were thirty-seven designed and engineered refineries running or under construction. There were also more than one hundred tea kettle operations consisting of a few connected oil drums that mainly skimmed the parts of the crude that produced gasoline of questionable quality.

As was widely known at the time, many of the more primitive refineries made money because their owners simply stole oil from large producers. Trying to protect their investments, these larger companies hired men to walk the length of their pipelines, looking for illegal taps as well as leaks. Bill Measures, who worked for Humble, discovered a one-inch tap that ended at one of the larger illegal operations. Ordinarily, companies like Humble reported illegal operations to local lawmen and waited for them to arrest the thieves. Humble, however, had learned that the thieves were often better connected in local politics than the wealthy corporation and that local officials rarely cracked down on oil thieves. So, when Measures reported the tap, Humble's managers took their own kind of justice by pumping cement through the illegal pipelines and into the refinery. Its retaliation did not end theft in the East Texas Field, but it apparently discouraged additional taps into Humble's lines.

Lawlessness in East Texas went far beyond oil theft. As had been true of earlier booming regions, the new field attracted gamblers, prostitutes, and moonshiners. More seriously, armed thugs carried out violent robberies and professional gunmen enforced the territorial claims of rumrunners and pimps. Lawmen in Henderson and Longview, both settlements really no more than dirt-street villages before the boom, struggled to keep a lid on local crime and succeeded in doing so more often than not. In the new tent cities, places like New Town and Pistol Hill, law was an abstraction, at best an obstacle to the tangible business of separating oil field workers from their hard-earned dollars. As lawlessness mounted and citizen complaints reached Austin, the familiar cry was heard— send in the Texas Rangers!

In this instance, the Rangers' presence was embodied by one of the more remarkable peacekeepers in their history—Manuel T. "Lone Wolf" Gonzaullas. With a flair for drama that must have impressed even J. Edgar Hoover, Gonzaullas slipped into Kilgore, unshaved and unwashed, dressed like an unemployed oil field hand. He scouted the towns and camps undercover, before he made his dramatic entrance—clean shaven, bronzed, brawny, equipped with a repeating rifle and two pearl-handled pistols, topped by a white Stetson hat, riding on a larger-than-life

Above: M. T. Gonzaullas, the legendary law in East Texas.

COURTESY OF THE EAST TEXAS OIL MUSEUM, KILGORE COLLEGE.

Top, right: A Texas National Guardsman watches a fire in the East Texas Field.

COURTESY OF THE PERMIAN BASIN PETROLEUM MUSEUM.

horse. His careful staging was enhanced by widespread rumors to the effect that Lone Wolf was bullet-scarred and had many, many notches on his guns but there was room for no more. "One riot—one Ranger!"

Actually, Austin had sent more than a dozen lawmen to carry out Lone Wolf's mission. In a well-planned campaign, they arrested nearly one hundred people a day over a three-week period, chaining many of them inside Kilgore's Baptist church, where they fed them once a day and passed a tin can to serve as a urinal. The arrest total might well have passed two thousand, not only beyond the capacity of the town jail, but far in excess of the number who could be tried in local courts. The solution to the potential judicial logjam was a tried-and-true Ranger tactic: after they had suffered for a few days, the relatively minor criminals were freed, on the condition that they clear town by sundown. Ring leaders, armed robbers, and murderers were held over for grand juries. Bootleggers, pimps, gamblers, and prostitutes were rarely arrested by the Rangers on the grounds that they represented a relatively minor threat to public order. In East Texas, that was a good call. After a month in the field, the Rangers were less visible, but the threat of another crackdown was often sufficient to support the

authority of local lawmen. The Rangers— more likely to ride in fast and well-armed cars than on Arabian stallions—had restored a semblance of order with selective enforcement, on a smaller scale in Mexia, Borger, and Wink, and the same approach worked again in East Texas..

The challenge mastered by the Rangers was, however, modest compared to the havoc wrought in the oil industry by the mounting wave of East Texas oil. The problem was easily identified: the field produced more crude oil than consumer demand for refined products warranted, so as the wells flowed, prices plunged. With them went the value of the assets of the large companies that held millions of barrels of crude and gasoline in tanks, and the worth of anyone, independent or large producer, who owned oil still in the ground. Though economic theory seemed to prove that supply and demand would balance, at least in the long run, by late 1931, leaders in the industry and in Texas government doubted that there would be many survivors at the end of that long run. The Rangers had the advantage of highly visible authority, based on laws that were respected—if not observed—by most Texans. It was much more difficult for other public officials to stem the tide of East Texas oil because there were no enabling laws—and there was no consensus among producers about possible remedies.

CHAPTER V

WHO WILL MAKE THE RULES?

"On reporting an oil sale at two-cents a barrel: "probably the lowest price ever paid for oil thus far in the twentieth century."
- *The Oil and Gas Journal*, July 10, 1931.

As government and industry tried to cope with the flood of East Texas oil and the economic ruin it inflicted on the oil industry, an otherwise obscure Texas agency emerged as one of the most powerful state regulatory bodies in American history. That entity, the Texas Railroad Commission, consisted of three members, elected for six-year terms. The legislature empowered the TRC to set intrastate railroad rates and supervise the delivery of rail services to Texas communities in 1891. Not a great innovation at a time when other states were taking similar steps, the commission was the Texas legislature's response to the populists' anti-railroad campaign, which was based on the belief of Texas farmers that railroads were exploiting them by charging unreasonable rates to carry farm products to markets. In fact, the political agitation and legislative response amounted to political scapegoating of the nearly bankrupt railroads for the market-driven ups and downs of agribusiness. In actual operation, the TRC could do little for farmers, but it did tilt the scales in favor of agribusiness processors in Texas, giving them competitive advantages over those elsewhere. Thus, under TRC rates, it was cheaper to send cotton bales to the Gulf Coast for processing than it was to move them over local lines that connected with interstate carriers.

For the first few decades of its existence, the commission was caught up in the political and mathematical complexities of setting rates for cotton from Wichita Falls to Galveston, and for wheat from the Panhandle to Dallas and Fort Worth, along with enforcing the continuation of rail service to all towns on the lines, even after local demand had declined to the point at which the railroads could

A mass meeting in Kilgore in protest of proration rules.
COURTESY OF THE EAST TEXAS OIL MUSEUM, KILGORE COLLEGE.

show only losses when they provided that service. It is clear, thus, from its beginning, the mission of the commission was political and economic: to deliver benefits to the politically active groups that took an interest in its proceedings—and in the election of commissioners. Had it done no more, the Railroad Commission would still be worthy of mention in Texas history textbooks.

The legal mission of the commission broadened over time, a common situation for regulatory bodies. The first major change occurred in 1917 when the legislature regulated pipelines and gave responsibility for enforcement to the commission on the grounds that it was a transportation issue and the commission was already involved in transportation. Logical enough, and frugal. The alternative was creation of an additional agency, which would have added to the state's budget. It was far cheaper to heap on responsibilities without adding revenues sufficient for responsible enforcement—an approach the Texas legislature long took with the TRC. Apart from obvious frugality, the legislature thus guaranteed that the commission would not intrude greatly on business unless it responded to public cries for action. Understaffed and underfunded, the TRC was often unable to compel even minimal compliance with its rules. In North Texas, for example, it discovered that a majority of the wells drilled in one highly active county were never covered by paperwork the operators were required to file with the TRC.

And so it went, as the Texas legislature continued to add to the TRC's responsibilities, giving it authority to prevent the physical waste of petroleum in 1919, thereby passing up the option of creating a conservation commission that might have had at least half a chance to enforce the relevant laws. Again, the reasoning in the legislature was that the TRC was already involved in the oil industry so it would be more efficient, and cheaper, to expand its responsibilities. Meager appropriations guaranteed that the commission's role would be largely limited to following up on complaints lodged by one oil producer against another, a limitation of its functions that lasted into the post-World War II period and beyond. Whatever nineteenth-century reformers thought they were creating, the TRC was, early on, largely involved in keeping peace within the oil industry; it was not an enforcer of what populists and their successors called "the public interest." In Texas, the public rarely knew who the commissioners were or what they did, a highly desirable situation as far as some commissioners were concerned.

So empowered, the commission held hearings and created rules that applied to the field operations of the oil industry. Some codified older regulations such as requiring casing in wells to protect ground water, a problem that was especially urgent during the development of the fields in central Texas. Even before 1919, the commission had tried without success to restrict drilling on small tracts, like the doormat-sized leases in North Texas fields. With new authority, the commission issued Rule 37 to accomplish that objective. Rule 37, the most controversial— and the one to which the commission granted thousands of exceptions—barred drilling within 150 feet of a property line or within 300 feet of another well. With the law of capture still recognized by Texas courts, there was little more the commission could do to impede the rush to drill that followed the discovery of oil fields.

When Rule 37 proved to be inadequate, the commission tried other strategies to curtail rapid drilling. In the Panhandle, for example, it tried to stop the "shooting" of oil wells to increase production. (See Chapter Seven for a description of well-shooting.) In line with what seemed to be the commission's mandate to conserve oil, it reasoned that during periods of flush production, the practice was unnecessary and wasteful. The problem with this interpretation of Texas law was that the TRC was empowered to stop both the physical waste of oil and damage to waterways from runoffs of spilled oil or leaks from the huge earthen storage pits that were used in many fields, but shooting wells did not necessarily waste oil even though it made it less economical to produce it when the practice led to more oil produced in limited or glutted markets, like the Panhandle. In short, in response to emerging problems, the commission attempted to expand its powers beyond those granted by the legislature, at the urging of some oilmen.

Such was the case in the Permian Basin during the late 1920s, as fields came to produce much more oil than the markets could absorb or than pipelines could carry to distant refineries. In both the Hendrick and Yates Fields, the commission was called in by producers to oversee the voluntary restriction of production. It took more time for many oilmen in the Hendrick Field to negotiate a method of allocating production than it did for the handful of producers in Yates, but both fields were choked back from maximum production by 1928 when the Railroad Commission acted as umpire to enforce the producers' agreements. Both pacts were based on formulas that included the maximum potential production from a well and the number of acres a producer had in the lease. Both factors had to be included, because omitting credit for the size of leases only encouraged oilmen to drill all of them up close together, adding to the problem of excess production. Proration, allocation of production short of the maximum possible for a well, was worked out by the producers themselves in that region, a situation that was clearly not possible in East Texas after "Dad" Joiner got lucky.

The list of producers in East Texas was almost as long as the Tyler phonebook, mainly because independents rushed in where majors feared to tread heavily. Like most independents, those in East Texas usually operated on a shoestring, borrowing, buying supplies on credit, and looking for as rapid a return on their investment as possible. For many of them, like Ed Bateman, "the long run" sometimes went no farther as the next payroll. Because of the added incentive to drill quickly and densely provided by the rule of capture, with only a handful of exceptions—including H. L. Hunt and Robert R. Penn—independents were strongly opposed to attempts to limit production from the flush field. They were also more than ready to use their local political clout to command the attention of legislators, and they were just as

ready to file suits in state and federal courts to defend their interests. And so they did, whenever the Railroad Commission or any other part of Texas government got in the way.

Thus, when Governor Ross Sterling called the Texas legislature into special session to address the problem of excess production in East Texas, before that body could act, a federal court ruled that the state could not fix prices by regulating production. With no statutory option, the governor sent the National Guard and the Texas Rangers into the field to shut down more than sixteen hundred flowing wells, under the guise of maintaining public order. In the face of Sterling's show of force, producers continued to ship "hot oil" and local refiners stayed in business. Then, in February 1932, another federal court order ended the role of the Guard and Rangers. Prices plummeted once again, until the Texas Railroad Commission more or less shut down the field for three weeks in April, claiming that some time was needed to gather data and develop a remedy. By May, however, oil was back to a dime a barrel and Texas had run out of legal remedies.

At the urging of Texas congressmen, President Franklin D. Roosevelt stepped into the scene in mid-1933, sending federal inspectors to gather data on production and

Ernest O. Thompson, Texas Railroad Commission member (center) and Olin Culberson, TRC chairman (right) with M. A. Machris and Frank W. Lake at a Wilshire Oil Company dinner in Austin.

refining, hoping to keep production in line with Railroad Commission goals. For nearly eight months, production stabilized, until the United States Supreme Court struck down Roosevelt's legal authorization to intervene. It was back to square one, but not for long. Moving with exceptional speed, Congress passed the Hot Oil Act in 1935, making it illegal to ship "hot oil" over state lines when it was produced in excess of the volume permitted by state regulatory authorities. The Texas legislature followed up with new legislation authorizing the confiscation of hot oil. After four years of massive production and hardship prices, producers in the giant field were limited to seventeen producing days and as little as three percent of a well's potential.

During the turmoil, Ernest O. Thompson, former mayor of Amarillo, emerged as the Railroad Commission's leader and public spokesman. Thompson was plain-spoken and well-versed in the complexities of the oil industry. For three decades, he met the press and rallied legislative and public support for the regulatory body. In 1940, he was joined by Olin Culberson, who served three six-year terms. Culberson, expert in the complexities of utilities regulation, added talent and credibility to the body. He and Thompson also cooperated to replace some of the political appointees on the TRC staff with trained petroleum engineers, further enhancing the body's effectiveness, though field staffs were still small and overextended. As notably, Culberson worked to make the Commission's work more accessible to the public by scheduling weekly meetings and requiring a ten-day advance notice of pending hearings. In disputes over prorationing, Culberson was usually known as the defender of the Texas independent oilmen. Between them, Thompson and Culberson brought the TRC out of the comfortable obscurity their predecessors had enjoyed and engaged more fully in the political life of the body. Though generally supportive of measures on which the oil industry agreed, they both used occasions on which there was no unanimity to enhance the TRC's role as peacekeeper in the often contentious oil industry.

By the end of the Depression, an otherwise obscure regulatory body had emerged as critically important to the petroleum industry, and not only in that of Texas. When additional discoveries were made in future years, the Commission's enforcement of proration and of shut-down days supported the price of crude oil by matching supply to demand. As industry observers would note forty years later on the East Texas crisis, the TRC provided the model for the Oil Producing and Exporting Countries (OPEC) when it aspired to the same power over markets. Even then, the three elected commissioners were more empowered by industry participants who pursued public policies in their own interests than they were by either the electorate or the legislature. It was always highly unlikely that the body would either create a rule or enforce a law that lacked strong support in all segments of the oil industry.

The genius of the proration system is found in its yielding desirable consequences for most industry participants. The refineries provided demand figures for the TRC, so they were assured of feedstocks at low prices. In the oil fields, proration formulas protected the investment of large companies in reserves by supporting prices; at the same time, proration produced a more predictable price for produced oil, an advantage for all producers. As long as the TRC was willing to make exceptions to its rules, which it was generally willing to do unless other producers objected to requested waivers, the work of the commissioners went unchallenged and they remained re-electable.

In light of the compliance of the TRC with policies and practices supported by the industry it regulated, it is still notable that its authority increased overtime. Apart from the East Texas crisis, it is unlikely that Texas, with its deeply individualistic political culture and well-organized economic interest groups, would have created a super-agency that would seem to tell businessmen how to use their property. Although TRC's authority was often far less than one would expect, on the basis of its statutory authority, it is still notable that it could ever restrict any part of the oil industry, let alone do it for more than three decades.

CHAPTER VI

WHILE EAST TEXAS GRABBED THE HEADLINES...

"You couldn't stand a banker on his head on a Main Street corner and shake $1,000 out of him."
- George W. Strake on raising venture capital during the Depression.

Had economics and regulation determined the course of history, exploration would have stopped completely when the tidal wave of Texas oil hit the market. With prices only a small fraction of the cost of drilling to replace the oil, there was scant motivation for oilmen to take on the unavoidably high risks that came with the search for oil on geological frontiers. Proration meant that new discoveries would take much longer to pay for themselves and, as always, few independents could afford to take a long view of energy economics. Bills came due today if not yesterday. Economics and regulation thus made risks much higher. Even the large integrated companies had serious problems. Despite the bargain-basement prices they paid for crude oil, they were not doing especially well as a group because the worldwide depression curtailed demand for refined products. Among the major companies, only Humble, then a subsidiary of Exxon, continued to upgrade its refineries during the decade, though Texaco and Shell would make significant improvements near the end of the decade. Reasonably, one would have expected exploration and plant expansion to come to a grinding halt during the 1930s. That is, if one forgets the ingenuity and tenacity of oilmen.

Exploration was driven by a number of changes, mainly improvements in geophysics and the accumulated scientific data that indicated high probabilities of commercial oil production in most parts of Texas. Some of this data was already in-hand when the East Texas Field became the focus of investment and activity. Once production from that field and others was firmly controlled by the Texas Railroad Commission, prices for crude oil rose, making the saved data immediately relevant. The

A gas blowout leveled a derrick at Refugio in 1931.

Pioneer geoscientist Mrs. H. H. Adams in the field.
COURTESY OF THE SOUTHWEST COLLECTION, TEXAS TECH UNIVERSITY.

aggressive expansion of its reserves, which it did, beginning with the East Texas Field, in part, because its earlier investments in pipeline and plants continued to produce services and products at highly competitive prices. The company continued to show sustained profits even during the Depression years.

By paying meager dividends, Humble retained earnings to purchase and explore for reserves, which increased from 400 million barrels in 1930 to more than 2.5 billion barrels of oil and 6 trillion cubic feet of gas by the end of the decade. Of even greater significance for the future of the company, the lion's share of reserves added during the latter half of the 1920s had been located by application of increasingly effective geophysical procedures. Going into the Depression and the East Texas crisis, Humble had a large backlog of data, which it continued to supplement. During slow times, in 1932, for example, Humble did more than forty percent of all geophysical field work in Texas. During the same year, Shell retrenched to pay its loans, Gulf continued to lose money, and Texaco redirected its investments to retail operations. At Humble, the chief geologist, Wallace Pratt, and the company's CEO, Will Farish, focused on reserves to meet their refinery needs and those of Exxon, and to take advantage of the fact that there was little competition for good leases at the time.

Humble's aggressive reserves acquisition strategy was firmly in place even before "Dad" Joiner's well came in. As its oil scouts, along with those of other companies, followed the progress of the well test, Humble leased twelve thousand acres. When Ed Bateman needed money to develop his discovery, Humble paid his asking price, $2.1 million for 1,500 acres. And Humble's landmen kept leasing, adding ten thousand more acres within two years. By the time it had ended its acquisition campaign in the East Texas Field, the company had added six hundred million barrels of reserves, more than doubling its previous holdings.

Steadily, Humble bought reserves wherever oil was found in commercial quantities. When Hugh Roy Cullen and West Production brought in the Thompson Field in 1931, Humble bought half of their holdings. After

curtailment of the field also encouraged the large integrated companies to find or purchase additional reserves because the great field could no longer meet all of their refinery needs.

In 1931, however, only Humble was in a position to take costly new initiatives. During that year, Gulf lost $23 million and could not borrow more money for new plays. Shell had an even larger deficit of $27 million, largely because extensive pipeline construction was done with borrowed money and debt service was expensive. Amoco and Chevron were hard hit by sharp reductions in gasoline sales and they avoided new capital commitments. Of the smaller companies, Marathon, Continental, Sinclair, and Phillips were awash in red ink. Humble, having paid off debts incurred to expand its Bayside refineries and pipeline systems during the 1920s, undertook

George Strake, a Houston independent, drilled a poor-boy discovery well and a confirmation well on ninety-three hundred acres in the Conroe Field in 1932, Humble hustled and bought leases overlying half of what would become that giant field's productive acreage, adding three hundred million barrels of reserves. The following year, Humble made its biggest leasing coup, picking up one million acres on the giant King Ranch, the biggest private lease transaction in American history to date. The company continued its leasing campaign in Southwest Texas, eventually acquiring two million acres between Corpus Christi and the border with the Republic of Mexico. The following year, it bought half of Cullen's leases in the new Sugarland Field, acquiring 75 percent of the productive acreage and adding 150 million barrels to its reserves. Even in a troubled industry, there was money to be made by those like Humble who had money—or at least daring and imagination, like Strake and Cullen.

Thus, even while East Texas oil flooded markets, wildcatters drilled. George Strake's discovery at Conroe stimulated exploration in the upper Gulf Coast area, leading to the opening of the Tomball Field near Houston in 1933, the Hastings Field in Brazoria County in 1934, the Tom O'Connor Field in Refugio County, and the Webster Field near Houston two years later. Humble's exploration programs continued to enjoy success during the rest of the 1930s, with its own discoveries in the Gulf Coast region at Anahuac and Friendswood. Together, these fields brought the company six hundred million barrels of reserves. Humble also acquired vast leases over the Katy Field, which produced huge amounts of wet gas, which was stripped to raise the octane rating of gasoline. When East Texas ignited again with the discovery of the Talco Field in 1936 and the Hawkins Field four years later, the company bought aggressively, adding more than 350 million barrels to reserves. As the decade ended in 1940, Humble spent more retained earnings to lease more than three million additional acres, about fifteen percent of them in what would become prolific producing areas in the Permian Basin.

Though Humble's success is certainly the major news among large companies during

Boom-time housing north of Odessa.

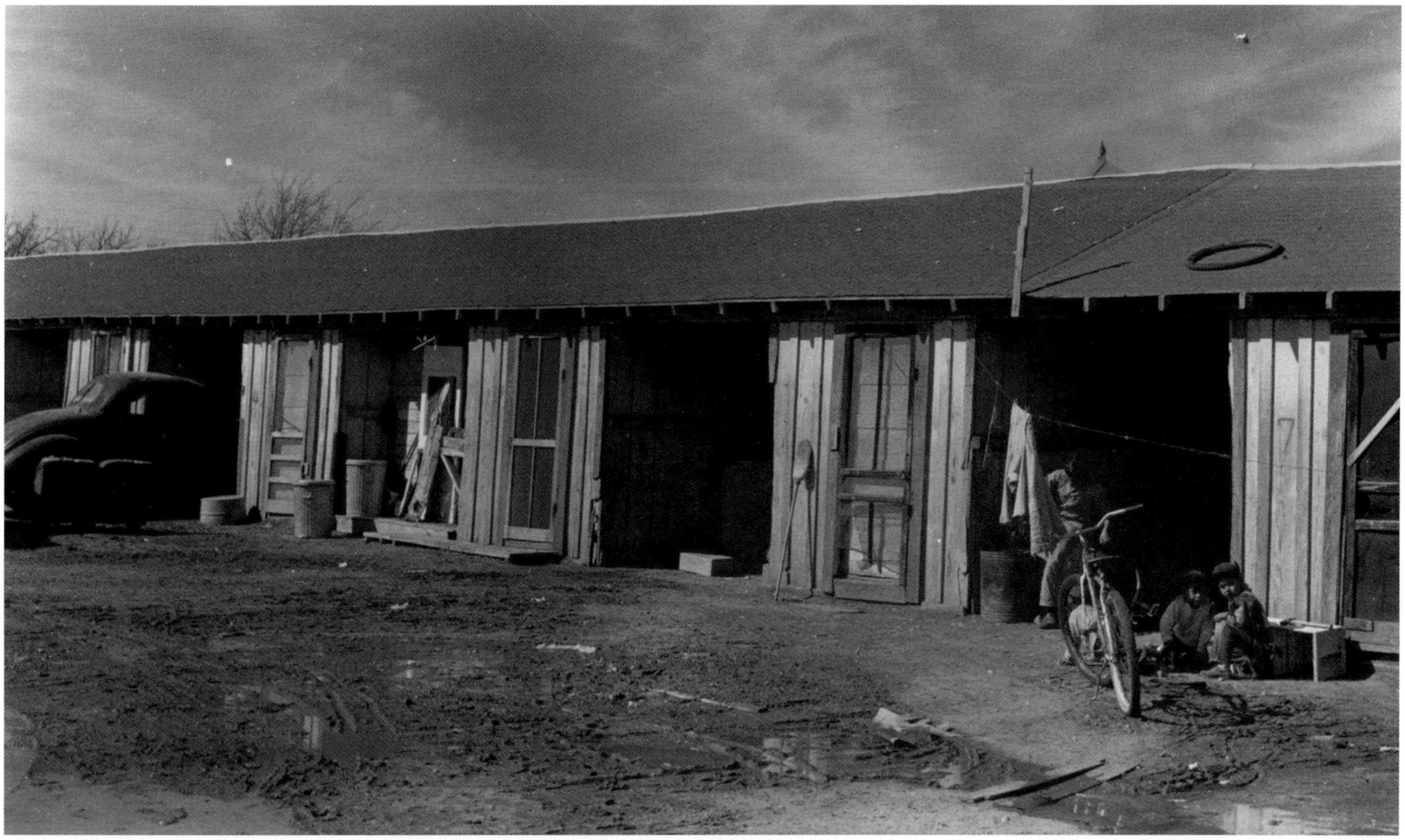

Sid W. Richardson, the man who was lucky and smart.

seismography. When University 2-B came in at twenty thousand bopd, oilmen knew that beyond Yates and Hendrick, the Basin and the plains north of it were prime targets.

During the first half of the decade, however, except for Humble, most large companies could not afford to drill more wells, especially while crude oil markets were glutted. As years passed, the large companies faced an urgent problem: they had leased millions of acres that they were not exploring, and their leases commonly ran for ten years. Unable to drill the leases themselves, the companies turned to independents, offering them a range of incentives to take the associated risks. The biggest lure was the transfer or farming-out of leases from large companies to the independents. Commonly, the independents sold part of that acreage to pay costs of drilling and improving the leases. Beyond that support, independents often received dry-hole money, if they drilled to the agreed-up depth without success, and bottom-hole money when the well was drilled successfully. None of these contributions paid the full cost of completing a well, but they lowered the amount of capital the independent risked from his own pocket and those of his investors.

With farmed-out leases in hand, the typical independent then sought to fulfill his obligations as cheaply as possible, cutting corners by buying used pipe, doing the management and supervision himself, by trading some of his interest in the well for services and supplies, and by hiring workers who were flexible about the arrival of payday. Working lean, if not mean, was called "poor-boying" and it was pervasive in the oil fields. Workers later said that when money was really tight, they were paid with groceries, which the independent had usually obtained on credit at a neighboring store. In the oil fields, these positions were called "bean jobs."

One of the most successful "poor boy" independents, Sid W. Richardson, did not remain poor long. By 1930, Richardson had made and lost two fortunes. Then opportunity knocked when large companies needed to drill to hold leases and gave Richardson farm-outs on leases that were not

the Depression, during the second half of the 1930s, its large competitors came back in the game. In the Permian Basin in particular, Shell, Amoco, Texaco, Mobil, Chevron, and Gulf made significant finds and acquisitions in a region that yielded its secrets with increasing frequency to seismologists who exploded dynamite and measured and interpreted shock waves in more than a dozen counties. There was sound judgment behind expensive investments in geoscience. Following the completion of University 1-B in the Big Lake Field during 1928, the deepest well in the world when completed at 8,526 feet, explorers knew that deeper drilling in the area would probably identify producing horizons that were not yet easily located with

considered rank wildcat because they were either in or near producing leases, or were supported by what seemed like good geological data. Richardson, a capable and honest driller even when he was near penniless, drilled on major company leases in the Kermit and Keystone Fields early in the decade. Thereafter, he would never be poor again, emerging as one of America's wealthiest businessmen by the 1940s.

George T. Abell, another independent, quit his job with a large company to go independent when his employer cut back on exploration. With a farm-out from it in hand, Abell borrowed money against his home, his car, and from a loan shark to make a small find near the Pecos River. He sold his interest in that lease, rolled the money into a farm-out from Mobil, which committed bottom-hole money along with other leaseholders, and completed a number of wells that produced between 1,200 and 2,400 barrels per day before the field was subject to TRC proration. Not surprisingly, the field was named after him. It was the making of his fortune, which he went on to increase in the oil fields of West Texas.

On the heels of discovery of the North and South Cowden and Goldsmith Fields near Odessa, and the Fuhrman and Means Fields to the north, Humble would pay for reflection seismograph work on thirty-one sections in Andrews County. North of that area,

Amerada, managed by Everett DeGolyer, a pioneer in geology and geophysics, shot more leases and found significant production near Seminole. Conoco farmed out leases to Amon Carter, a Fort Worth publisher and owner of producing and drilling companies, who brought in the Wasson Field. Texas Pacific Coal and Oil and Conoco gave farm-outs to C. J. "Red" Davidson, another Fort Worth operator, who brought in the Bennett Field. And so it went through the vast region as independents drilled with farm-outs, bottom-hole, and dry-hole money, proving acreage that stretched more than 100 miles north of Odessa and nearly 200 miles from the Big Lake discovery to open the region to extensive exploration.

Even during adverse times, new fields were discovered, nearly doubling the producible reserves of Texas. When times improved, work in the region expanded, with more than four hundred wells completed in just the North Cowden Field between 1936 and 1940. Another big find in the area came with the discovery of the Slaughter Field in 1936. Development of the field occurred intermittently during the next five decades, until Amoco, Sid Richardson, and other operators had drilled 2,535 producing wells in it. With the succession of new discoveries during the 1930s, the Permian Basin was known as the locale of major reserves and as the potential arena for even more discoveries.

Goldsmith, a boomtown in the making north of Odessa, in 1937.

Both Slaughter and Wasson Fields would be further developed during the 1960s and 1970s through massive water flood and other enhanced oil recovery programs. By the end of the twentieth century, Wasson had produced about two billion barrels of oil, making it second only to the East Texas Field. Cumulative production in the Slaughter Field passed one billion barrels in the early 1990s.

Southwest Texas was slower to revive than the Permian Basin-Plains area, but several notable discoveries encouraged future exploration in the region. In the Saxtet Field, near Corpus Christi, oilmen drilled below the gas-producing formation and identified thirty additional producing horizons. Mobil's discovery of the vast Seeligson Field in 1937 confirmed the multi-pay potential of the region as the field ultimately produced from 150 separate reservoirs. When the Railroad Commission established separate quotas for the producing sands in the Greta Field, it encouraged additional drilling in the area. Producers then reentered old wells to take advantage of the multiple horizon find at Saxtet and the new commission policy. With the opening of the Benavides, Tom O'Connor, LaGloria, Placedo, Seven Sisters, Corpus Christi, and other fields, southwest Texas emerged as a major supplier of natural gas for the cities and industries of Texas and as a significant oil province. The region had become so important by 1937 that Humble built the first sizeable refinery, near Corpus Christi, at Ingleside. Southwest Texas had arrived.

By 1940, Texas was the leading American producer of both oil and gas, supplying nearly a trillion cubic feet of gas and half a billion barrels of crude oil annually. There was also vast potential production that was not utilized because of statewide proration of oil and the open-air flaring of natural gas in many fields. The annual value of oil and gas produced had passed the state's income from cotton more than a decade earlier. By the end of the 1930s, Texas was clearly the major producing state in the nation, the Railroad Commission was deciding how much oil each well could produce, and the largest concentration of refineries in the world had been built and expanded along the upper Gulf coast of the state.

With more than a quarter of a million Texans employed in the industry, with good wages and salaries—compared to the income of cotton farmers—the industry had become the keystone in the state's economy. As production increased, so did revenue from taxes on oil production, so that nearly one-third of the cost of state government—and even more of the expense of many local governments and school districts—were met by levies on oil production and properties. During the four decades that followed the great discovery at Beaumont, "Texas" came to mean "oil."

Goldsmith, Texas.

CHAPTER **VII**

HOW THE WORK WAS DONE

"They came from everywhere! And they was as good as the land they stood on."
- Bill Measures on boomers and drifters.

Texans, by the hundreds of thousands, were actively employed in the petroleum industry by the 1930s. "Upstream," in exploration and production, they worked in geological and seismic crews, moved dirt, handled bits and pipe on drilling rigs, shot or acidized wells, laid pipelines, operated them, and built storage tanks. "Down stream," men and women earned their livings in refineries, wholesale agencies, in management centers, and at filling stations. Most of the upstream work was done out-of-doors, in all kinds of weather, and it was dangerous because explosions, fires, collapsing rigs, falling pipe, and other hazards were common occurrences. Inside the refinery gates, explosion, fire, and toxic gasses threatened lives, risks that were also common in the oil fields. Most workers seem to have believed that the benefits of working in the petroleum industry

Rig builders on break near Humble, 1907.
COURTESY OF THE PAUL SILL COLLECTION,
PERMIAN MUSEUM PETROLEUM MUSEUM.

Above: Special rigs like this Fort Worth spudder were often used to start drilling.

COURTESY OF THE PERMIAN BASIN PETROLEUM MUSEUM.

Below: Teamsters moving a rig in the Burkburnett Field in 1918.

PHOTO BY CLYDE BARTLETT, COURTESY OF THE
C. C. RISTER COLLECTION, SOUTHWEST COLLECTION,
TEXAS TECH UNIVERSITY.

workers guessed that the mules liked the songs that black teamsters shouted and exchanged. Perhaps that was so. Cowboys had long believed that even off-key singing calmed herds. For a number of reasons, including the fact that teaming was one of the lowest paid jobs in the oil fields, black men often did the dirt work and hauled timber and pipe to rigs.

Once a well site was leveled and pre-cut timber was on the location, rig builders arrived to erect the wooden rig, long the most familiar site and symbol of the oil field. With saws, large axes, and oversized hammers, they crafted the rigs and the drilling apparatus that could be fashioned out of wood. Most of the time, they worked until sunset; even then, some oilmen installed electric lights after generators were available, to extend their workday because nothing else could be done until the rig was complete. The rig builders saw themselves as craftsmen and they took advantage of the demand for their services that came with every boom—to organize labor unions, and to secure higher pay, something few other workers in the oil fields managed to do. They were also respected in the oil fields for their ability to work and play hard. In Odessa, some of them met at a local pool hall after work to drink and fight. One pair, apparently close friends, swung fists at each other so often that the bartenders put their best liquor and anything else that was breakable on the floor behind the bar whenever the two rig builders walked through the doors.

outweighed the undeniable risks. Mainly, the industry offered employment, usually at higher rates along with higher risks, that was more rewarding than fighting nature and crop surpluses on cotton farms or sweating through long days, inhaling lint, at the gins.

The first oil workers on a scene were those who prepared drilling sites, using horse and mule teams to level locations for rigs. Along the Gulf Coast and in North and East Texas, the dirt workers were common African-American men, who were thought by some white workers to have a special knack of handling animals. Some of the young Anglo

Those of us who tend machines or work at desks might well wonder where they found energy to spar and punch at the end of a strenuous work day of lifting and hammering.

The specifications of the rig builders' projects varied somewhat, mainly according to the kind of drilling apparatus that would be installed. The most common in North Texas, for example, was the cable tool rig, which "made hole" when the weighted bit slammed into the earth. On these rigs, the builders constructed large bull wheels that lifted bits for their downward plunge. They skipped that step in construction when they erected rigs that used a rotary drilling method, which made holes with a twisting bit mounted on the bottom of the pipe.

The type of rig employed on a well site depended on the underlying minerals. When it was limestone, or another hard mineral, the cable tool rig made holes quickly and efficiently. However, when drill bits had to penetrate soft sand, as they had at Spindletop, rotary equipment was needed to keep the hole from filling in after each probe. The drilling crews also ran larger pipe, casing, to hold back sand and to seal off water.

It made a considerable difference as to whether a worker was on the floor of a cable tool rig or a rotary rig. The cable tool rig crew was usually made up of only two men, a driller and a tool dresser. The driller kept the bit working and tried to monitor whatever was going on down-hole by holding the drilling rope or cable, sniffing for gas, and examining

Above: A drilling crew in North Texas enclosed part of its rig to cut cold winds.
COURTESY OF THE PERMIAN BASIN PETROLEUM MUSEUM.

Below: Carl B. King, later a highly successful driller, on a rig in the 1920s.

Two of the roughnecks and the driller worked on the rig floor connecting and disconnecting pipe as the drill ground downward, while the third roughneck worked high inside the derrick, guiding pipe during the pulling and re-entry phases of work. They also tended to work twelve-hour shifts or tours (pronounced "towers") and stayed on the driller's payroll until the job was completed. If the well produced oil or large quantities of natural gas, the well was "completed" when control valves were installed and connected to pipes that either carried the oil into storage or the gas or oil into a small pipe that gathered oil and moved it along to larger trunk pipelines, which carried it to distant users. When the well was a dry hole, they often pulled both pipe and casing for salvage sale, then plugged the hole. Then, they moved on to new jobs if they had found them. If they were unable to find more work locally, they joined what they called "the suitcase parade."

Bessie Leonard married an oil field worker and joined the parade:

> When you moved, that was really a deal. You just tied stuff all over your car. The thing of it was, them days, it didn't seem like anyone had any more than the other guy. We lived at about the same standard of living. They were all looking for work. They would just load their car with what they could put inside, on top if it, on the running boards, tie the spare tires on somewhere and take off.

J. M. Horner, started work as a teamster in the West Virginia oil fields, went on to build rigs in Pennsylvania and held half a dozen different jobs in as many places until he moved to Oklahoma to dress tools on a cable tool rig. From there, he migrated to Mexico City to drill water wells, then back to Pennsylvania and West Virginia to drill more oil wells. After more than ten moves, Horner was just 28 years old. At that point in his life, he took the train to Ranger and worked as a driller before he moved along to Breckenridge, Eastland, Gorman, and to two Oklahoma fields. He arrived in the Permian Basin in 1926, working at McCamey until he settled down in Wink.

Most producing oil leases contained storage tanks before the Second World War.

Top: Drivers hauling nitro (in the box) and torpedoes near Houston.
PHOTO BY GRIGGS STUDIO. COURTESY OF
C. C. RISTER COLLECTION, SOUTHWEST COLLECTION,
TEXAS TECH UNIVERSITY.

Above: Pouring the soup in an old cable tool rig at Breckenridge, 1922.
PHOTO BY BASIL CLEMONS, C. C. RISTER
COLLECTION, TEXAS TECH UNIVERSITY.

bits or rock that came up in the bailer, which he used to clean out the hole from time to time. His assistant, the tool dresser, kept the drill bits sharp by heating them in a nearby fire and hammering them back to a point. Sometimes the tool dresser did double duty as fireman, keeping the power-generating boiler stoked. Both the driller and the tool dresser usually worked twelve hours a day, seven days a week, or until they either completed or abandoned a well.

On rotary rigs, crews were larger including at least four men, the driller and three roughnecks.

At Spindletop, for two decades thereafter, the tanks were really over-sized barrels on end. Carpenters built them out of durable hardwoods then bound them with iron rings or cable. Beginning in the 1920s, metal tanks replaced the durable wooden type, bringing metal workers into the oil fields to construct, rivet and test them. Some producers preferred the wooden tanks because they resisted corrosion from chemicals better than early metal tanks. Over time, the metal tanks were treated to make them resistant to corrosion from crude oil. During the 1920s, oilmen also installed treatment equipment on their leases, especially if the oil was heavily "cut" with other minerals or if "wet gas" was produced along with crude oil and could be separated and used to enrich gasoline and other refined products. By the 1930s, the "old-time" look of the oil patch had largely disappeared from new fields. In East Texas, metal workers built oil rigs while less skilled workers assembled metal tanks.

Even if a well produced moderately well, oilmen often tried to coax more crude out of the ground. From the early days of Pennsylvania oil onward, the most common way to stimulate production was to explode nitroglycerine down inside the well. Specialists did this work at great hazard. In North Texas, shots of four hundred quarts or more were not unusual. One well in Stephens County took one thousand quarts. The nitroglycerine was produced from sulphuric and nitric acid and glycerin oil, mainly, in large tubs lined with water-cooling coils. Once these ingredients were mixed, they were treated to remove the acids from the nitro. At that stage of the process, if the temperature of the chemical rose and workers failed to control it, they bolted out of the factory doors as fast as they could move. If they were fast and lucky, they survived to salvage equipment and set up operations again.

Once the "soup" was manufactured, workers poured it into canisters, and carried them to trucks, where they lowered the containers into rubber holders designed to resist leakage and jarring. The equipment was far from fail-safe. Dirt roads to oil fields were often sites of nitro disasters. The road that

Above: Loading canisters near Houston, 1920.

PHOTO BY GRIGGS STUDIO. COURTESY OF THE C. C. RISTER COLLECTION, SOUTHWEST COLLECTION, TEXAS TECH UNIVERSITY.

Below: Well-fire fighters at work north of Odessa, 1937.

COURTESY OF THE JOE W. GRAYBEAL SCRAPBOOK COLLECTION, PERMIAN BASIN PETROLEUM MUSEUM.

connected Wink with Monahans, in the Permian Basin, had at least one deep crater, created when a truck and its driver were destroyed in a noisy flash. It is not surprising that other workers rarely hitched rides with nitro trucks!

Once the nitro was delivered to a rig floor, the shooter and the driller agreed on the size of the shot. Then, the shooter and an assistant poured the "soup" from canisters into long

Top, left: Giving Mother Earth a stomach ache with two hundred quarts of nitro.
RICHARD DONNELLY COLLECTION PHOTO, PERMIAN BASIN PETROLEUM MUSEUM.

Top, right: Workers stand back to watch a 575-quart shot northwest of Odessa in 1937.
COURTESY OF JOE W. GRAYBEAL SCRAPBOOK COLLECTION PHOTO, PERMIAN BASIN PETROLEUM MUSEUM.

Right: Every worker's worst nightmare—a blowout starts a major fire at a rig.
COURTESY OF THE PERMIAN BASIN ARCHIVES, UT-PERMIAN BASIN.

tubes, called "torpedoes," careful not to spill any of the highly volatile chemical on either the rig floor or on the outside of the container. Next, they lowered the torpedoes down the well and set an explosive device, usually called "the bomb," on top. After World War I, one of the most popular bombs was labeled "The Bolshevik Special." Another popular detonator was called the "Zero Hour Time Bomb" because the device detonated when the timer's hand reached a double zero at the twelve o'clock position on its dial. Finally, the shooters cleared out and waited for the detonator to touch off the expected explosion. Sometimes the blasts did not occur as expected so the shooters had to take the calculated risk of going back to the wells and using new detonators, hoping that the first ones would not kick in while they worked.

Shooting wells was not only dangerous and occasionally difficult to control, it was also costly to complete because the resultant downhole explosion left debris to be cleaned out, work that often took longer than setting up the shot. For that reason, producers looked for alterative ways to stimulate production. During the late 1920s, Dow Chemical convinced Pure Oil and other companies to inject acids to accomplish the

same purpose. The new technology caught on—acidizing had replaced shooting by the 1950s. During that decade, Amoco and other companies began to pump pressurized water and other chemicals down wells to break up or "fracture" formations. The shooters who survived the perils of their work, lost it to the new acidizers and hydro-fracers. The manufacturer of the zero-hour device went on to make Zebco fishing tackle, by all accounts safer to use than the bombs.

While the work lasted, some shooters became celebrities because of the dangers they faced. Tex Thornton, for example, made

headlines, as early as 1929, when he extinguished a well fire near McCamey. Spewing 75 million cubic feet of natural gas per day, the well lighted up the surrounding countryside for three days, until Thornton blew it out with a 30-quart nitro charge. Thornton made good headlines: "Tex Thornton, One Man Fire Department, Wins War With Flaming Well". In more recent times, Red Adair, Boots & Coots, and other well fire fighters faced even greater challenges and earned more headlines.

Pipeline construction, long labor-intensive, was transformed by the application of new science and technology. From the time of Spindletop into the 1920s, pipelines were laid by roving crews of unskilled workers hired by the large integrated oil companies that operated the lines. Though production workers tended to be permanent employees and drilling crews often stayed together for more than a single well, the old pipeliners were drifters. As longtime Humble Pipeline employee Bill Measures remembered them, they often lacked identifiable names and were known by physical characteristics—-Blacky, Shorty, and Redd—or by someplace they claimed to have lived—Oklahoma being second only to Texas in popularity. Before the days of social security registration and other formalities tied to hiring, they kept their

ears to the ground and followed work. As Measures described them:

> They came from everywhere. Why if you was like I was, you just walked out there and asked that old boy for a job. You didn't have no examination or a lot of papers to sign, all that stuff. You'd find out about work from each other. Just pass the word along. That was all there was to it.

Pipeline work was both physically demanding and poorly paid. As L. E. Windham

Above: Humble pipeliners connecting and lowering pipe in West Texas. Note the large tongs in right-center of the photograph.
PHOTO BY EXXON. COURTESY OF THE PERMIAN BASIN PETROLEUM MUSEUM.

Below: A Texaco ditching and connection crew, 1914.
COURTESY OF THE *TEXACO STAR,* JULY 1914.

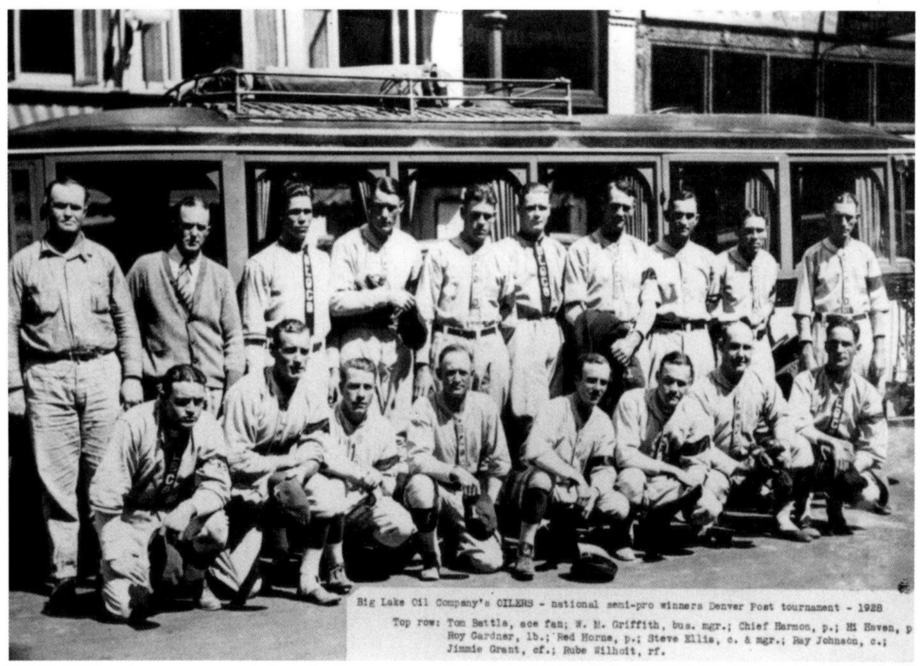

Above: Oil companies sponsored sports teams at many of their camps. One company, Producers (Texaco) bragged about its 1915 championship team.
COURTESY OF THE *TEXACO STAR*, JANUARY 1916.

Top, right: West Texas pipeliners lower tarred and wrapped pipe in the 1930s.
COURTESY OF THE J.A. MCVEAN COLLECTION, PERMIAN BASIN PETROLEUM MUSEUM.

Below: Sinclair's pipeline gang rounded up for field work in East Texas in the 1930s.
COURTESY OF THE EAST TEXAS OIL MUSEUM, KILGORE COLLEGE.

described it to S. D. Myres, more than thirty years ago, it included special skills and chores:

> You had a jack board man, you had a hook-man, you had a point man, you had a stabber to get it just exactly right to get the threads to match. They would take a pick handle and get on this end of the pipe and they could feel of it and know exactly when the threads matched. They would say "roll it" and the tong men would start pumping it, two men on each side, pumping the long tongs. They had a hammer man that beat the pipe like a drum beat, and they worked those tongs according to the hammer man. You had to because everybody had to be together or somebody would get hurt. You know those tongs were heavy.

Even before the pipe could be coupled, the route had been surveyed and ditches were dug to accept the pipe. Into the 1920s, most of the ditches were dug with pick and shovel, commonly by African-American men in the eastern half of Texas and by Tejanos in the southwestern part of the state. Once ditching machines came into common use during the later 1920s, crews that tended the machines could cut a mile a day in flat terrain territory. Often, several ditching crews worked on a line at the same time linking up at two or three points. Once the pipe was connected, workers swabbed it down with a protective chemical, lowered it, tested it, and buried it.

Pipeline work eased during the late 1920s when acetylene and electric welding were introduced to join solid pipe with more

reliable welds. With the change, the eight foot tongs disappeared from the oil fields along with the point men and the hammer men. Until the 1930s, most of the ditching and pipeline laying was done by company crews, housed in portable tent cities and fed out of company cook shacks. Thereafter, the larger oil companies found it more economical to let contracts for that work to smaller independent firms that specialized in pipeline work. They continued to hire informally and kept workers from contract to contract, with no expectation on either side that the jobs were more or less permanent.

Once the pipelines were in place, the larger trunk lines were equipped with compressor stations to maintain sufficient pressure to move the crude oil along. The stations were often as large as small factories, and they employed enough workers to bring Humble and other companies to build small rural company camps to attract and keep employees, along with their families, at isolated station sites.

In the upstream sector, men who kept the wells pumping and the pipelines working had regular jobs. Many of them spent the better part of their working lives working for just one or two companies. In their circles, there were career "Humble men," "Gulf men," "Texaco men," or "Marathon men." At the other end of the pipeline, refinery workers also had permanent jobs. The stillmen, firemen, plumbers, warehouse workers, and clerical employees could usually depend on regular paychecks. Pay varied considerably, according to skill and race, with African Americans commonly limited to warehouse and

maintenance work that paid less than process work. As a group, the refinery workers also had more leverage in terms of pay and working conditions, largely because they were much easier to locate than the floating oil field population so union organizers were relatively successful in recruiting them.

Even then, although organizers enjoyed some success from time to time, they usually lost ground during slowdowns at the plants. Some refiners, such as Humble, created company-sponsored unions to head-off organization by the stronger and militant trade unions that sought to enroll their workers. It was not until the 1930s, after federal laws made union organization easier, that powerful unions, such as the Oil Workers International Union, made permanent gains. Even then, permanent growth did not occur until the 1940s when the Oil Workers International Union won elections at the Gulf and Texaco refineries at Port Arthur, at the Amoco

refinery at Texas City, and at Shell's Pasadena plant. As membership continued to grow after World War II, the OWIU listed more than thirty-one thousand refinery and petrochemical installation employees in 1955. The number of unionized workers continued to increase along with the expansion of refineries and petrochemical installations, into the 1970s, when automation and other technological changes cut labor requirements and union memberships in those industries.

In the meantime, unions secured higher wages and improved benefits for their members, beyond those that had been available through company-sponsored unions. When management did not negotiate acceptable contracts, the unions shut down plants. For example, after VJ-Day, in 1945, the unions led a three-month strike that ended when the companies retracted plans for a substantial pay cut and agreed to raises that averaged ten percent. In following years, the unions also advanced the workers' interests in safety and other working conditions and tried to save jobs when new technologies and processes eliminated some positions. Even as union rolls shrank from the 1970s onward, the organizations were still powerful politically in the upper and lower Gulf Coast areas. The successor of the OWIU was smaller, but it was still able to push wage, safety, and job retention issues with management through decades marked by economic disruption and turmoil in international affairs.

Above: Texaco paleontologists comparing core samples in 1945.

Below: A young chemist monitors refinery production in the 1940s.

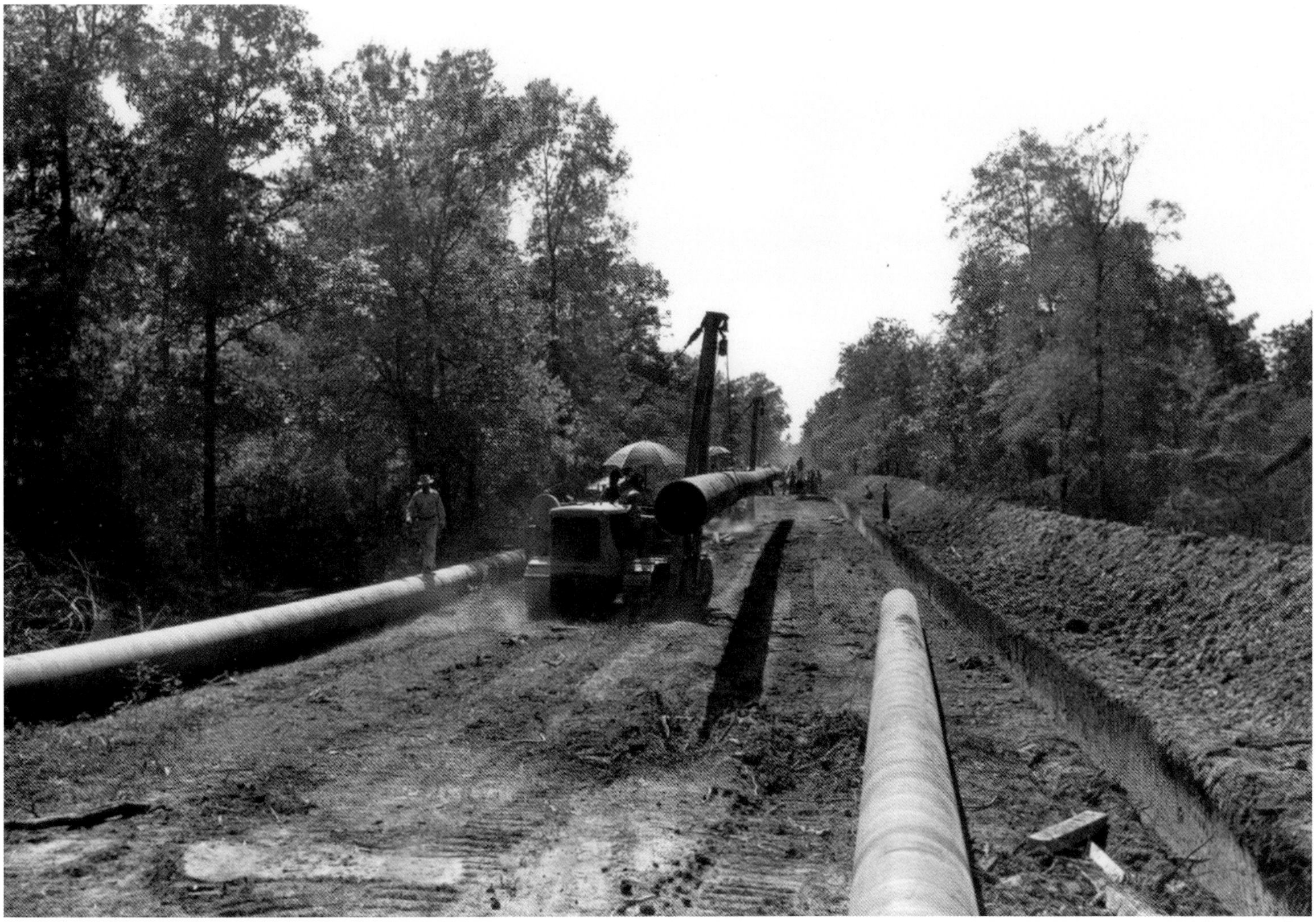

Chapter VIII

Oil and War

During the war years, Texans supported the crusade for freedom by serving in military forces and by working extra hard to meet urgent wartime demand for oil, gas, and refined products. Oilmen continued to drill wells, though hampered by shortages of workers and materials, and the industry expanded in yet another new direction as petrochemical installations multiplied near the Gulf Coast refineries.

Extensions of prewar exploration led to the discovery of two new fields, the Quitman in East Texas and Russell & North in the Permian Basin. Near Odessa, the most significant wartime discovery came in 1944 with the location of the prolific TXL Field, which would produce from a dozen different horizons in less than twenty years. There was nothing to match the discoveries of the 1920s and 1930s, however, because markets and regulation kept prices low and wartime materials allocation made it difficult to develop existing leases.

Before the United States entered World War II near the end of 1941, demand for refined products fell when France fell to Nazi invaders and the U.S. lost continental markets. Then, after America entered the war against Japan, Germany, and Italy, tanker traffic from the Gulf Coast to the East was interrupted by German submarines, which sank just enough tonnage to make marine insurance unaffordable by many shippers, diverting supply to railroad tank cars, inadequate in number and more

Building the "Big Inch" Pipeline.

costly to use. Demand for aviation fuel and gasoline increased, but the lack of efficient transportation choked back Texas production just when it was most needed elsewhere. In response, the federal government invested heavily in two massive pipeline projects, known as "Big Inch" and "Little Inch," which carried crude oil from Texas to the middle west, where it flowed eastward in existing pipelines.

The massive construction projects were completed in less than two years at a cost to the government of $146 million. Between them, the pipelines carried more than 350 million barrels of crude oil before the war ended in 1945. Thereafter, the government put the properties up for sale, accepting a bid from George and Herman Brown of Houston for $143 million. They went on to organize the Texas Eastern Transmission Corporation to operate the property and to make increasingly large investments in the oil industry.

Once the wartime pipeline system went on-line, Texas oil flowed freely with the East Texas Field permitted to produce wide-open for the first time. In fact, the field produced so much oil so rapidly that water incursion into producing wells speeded up in the field, bringing an increasing number of wells to produce more salt water than crude oil to the point that a significant number of them would be shut in, earlier than would have happened under proration.

Elsewhere in Texas, oilmen carried on as best they could in the face of shortages of skilled employees and quality materials, often fighting regulations that made little sense except to Harold Ickes, the new federal "petroleum czar". Experienced scientists were difficult to replace as were experienced hands. As a result, rig crews were increasingly made up of men too old to draft and school boys too young to serve in the armed forces. Many of them were willing and hard workers, but they were often less efficient than the men they replaced. Drilling and service work was more difficult because wartime materials priorities naturally favored the defense construction industry. The best steel went into guns, tanks and ships. Pipe and casing were always scarce, and the used goods available were often flawed, twisting off during drilling and required time-consuming "fishing jobs" to pull pipe and lost tools out of wells.

A new federal agency, the Petroleum Administration for War, headed by New Dealer Ickes, operated in five national districts. That measure of decentralization made sense given the size and spread of the oil industry. Difficulties arose, however, when the new bureaucracy created rules with the "one size fits all" mentality, leading to numerous requests for exceptions—all filed with at least two copies—from operators who believed that a rule was not reasonably applied to their operations. Conservation Order M-68, for example,

Night drilling near Odessa during the War, 1945.

resurrected a venerable though fallible conservationist notion that all oil wells should be drilled on wide spacing, at least forty acres. Reasonable as a general rule, the order interfered with the larger goal of increasing production in East Texas and other fields to meet current needs. In salt dome and other multiple pay fields, drilling on closer spacing was the most efficient way to produce additional oil and gas. Sometimes, given the geological complexity of producing formations, the rules failed to work as they were intended.

Federal oversight impeded the development of existing fields even when operators accepted the forty-acre spacing rule. Because pipe was allocated more readily for exploration than for development of existing fields, labor, and other material shortages made those wildcat wells less efficient to drill. Carl B. King, a veteran Dallas driller, for one, found that wells took almost three times as long to complete during the war as they had before it in the same oil fields. Inevitably, the materials-control systems produced massive amounts of paperwork and delays. One driller requested 700 pounds of arc-

welding electrodes, and got them, but only after he had completed nearly 200 priority forms, each submitted in triplicate and notarized. Under these circumstances, even as Everett DeGolyer and other oilmen moderated Ickes' regulatory zeal in Washington, it was difficult for oilmen to operate during the war.

Even when supplies could be obtained, they were relatively expensive because federal price controllers tended to let the cost of the materials that oilmen used rise more rapidly than oil and gas prices. The effect on operations was pervasive and predictable. Independents, in particular, were often hard-pressed to keep going. As veteran oilman W. D. Noel described it:

> We had a lot of wells shut in, the capacity shut in. And the manpower and the steel and the industrial equipment needed in drilling oil wells was needed in the war effort…. During this period of time the only priority we had was keeping the wells that we had producing…. We did sell one of our leases during that time; I think we would have had trouble servicing our debt had we not done that, because allowables

The Texaco refinery in Port Arthur, 1941.
The arrow points to the first boiler house,
constructed after Spindletop blew in.
COURTESY OF THE *TEXACO STAR*, OCTOBER 1941.

[in the Permian Basin] were so restricted. We did no drilling at all from the early part of '43…until the war was over.

Some of the most obviously obstructive federal rules were set aside quickly. One, for example, applied to all hauling, and required that trucks have cargos in both directions barring "dead-heading." Though the rule might have made sense as a means of conserving gasoline and rubber across the country, in the oil fields, it was a dead letter from the beginning, because it was obvious that a truck that hauled pipe to the oil fields could not bring back crude oil any more than the empty truck of a pumper could carry tubular goods on the outward bound leg of its journey.

In a more positive way, other federal initiatives boosted the Texas industry and moved it into what would become highly profitable new operations. The two big pipeline systems, "Big Inch" and "Little Inch," both operating by 1944, carried more than a half-million barrels of Texas crude northward every day, leading the Railroad Commission to increase production quotas in some fields. During the postwar period, utilities and manufacturers converted to natural gas; together with lines linking the Panhandle to the Middle West, these projects would transform the natural gas industry.

Of even greater importance for the future of Texas, wartime federal funds supported the development of a vast petrochemical industry in the state. During pre-war years, Gulf, Humble, Shell, Phillips, and other refiners had largely converted to cracking processes, refining crude oil under pressure and heat, with other chemicals, to increase the yield of gasoline from crude. The increasing abundance of "natural gasoline," the wet gas produced in the Katy and other fields, made the additional chemicals easier to obtain. The first gases, mainly butane and propane, yielded additives that raised the octane of gasoline to one hundred meeting the requirements of airplane engines and produced solvents such as methyl ethyl ketone. Little known, except to chemists, this product enhanced the flavor of margarine and the scent of perfumes; less glamorously, it yielded insecticides, paint and varnish removers,

adhesives, dewaxing compounds, patent leather, and printing ink.

Humble, Shell, and other refiners got into polymerization, the key petrochemical process, during the late 1930s. By the time the war began, both refiners were producing large volumes of aviation fuel. By the end of the war, petrochemical plants were tied to refineries with elaborate pipeline systems, producing vast amounts of basic petrochemical compounds, such as butane, propane, ethane, and naphtha; they got methane, the final basic building block in petrochemicals from natural gas separation processes. By then, the whole process had worked out into four stages. Oil and gas provided the feedstocks, including naphtha, methane, ethane, propane, and so on. The next step yielded basic commodities: methanol, ammonia, ethylene, propylene, and toluene, mainly. These base chemicals yielded additional chemicals and some final products such as formaldehyde, nitric acid, and ethylene dioxide. Additional processing produced such final products as resins, fertilizer, polyester, synthetic fibers, and polyurethane foam, though commercialization of many of these final stage chemicals would follow the end of hostilities.

Much of the petrochemical operation was given to the production of synthetic rubber, needed after Japanese troops seized what are now Malaysia and Indonesia. Trading some American science and gasoline enrichment formulas, Jersey Standard obtained German processes for creation of synthetic rubber from a German firm, I. G. Farben. Gulf Coast refiners provided the necessary chemicals to Firestone Tire and Rubber, B. F. Goodrich, and other producers of the new substitute for natural rubber. The addition of the rubber plants created still more markets for products of Texas refineries, and the production of synthetic rubber resulted in additional chemicals that were used by the developing petrochemical installations on the Gulf coast, bringing new plants built by Shell, DuPont, Dow, Celanese, and Monsanto.

By the time Japanese officials signed surrender documents on the deck of the U.S.S. *Missouri* in 1945, the petroleum industry in Texas had both grown and diversified, laying the foundations for postwar expansion.

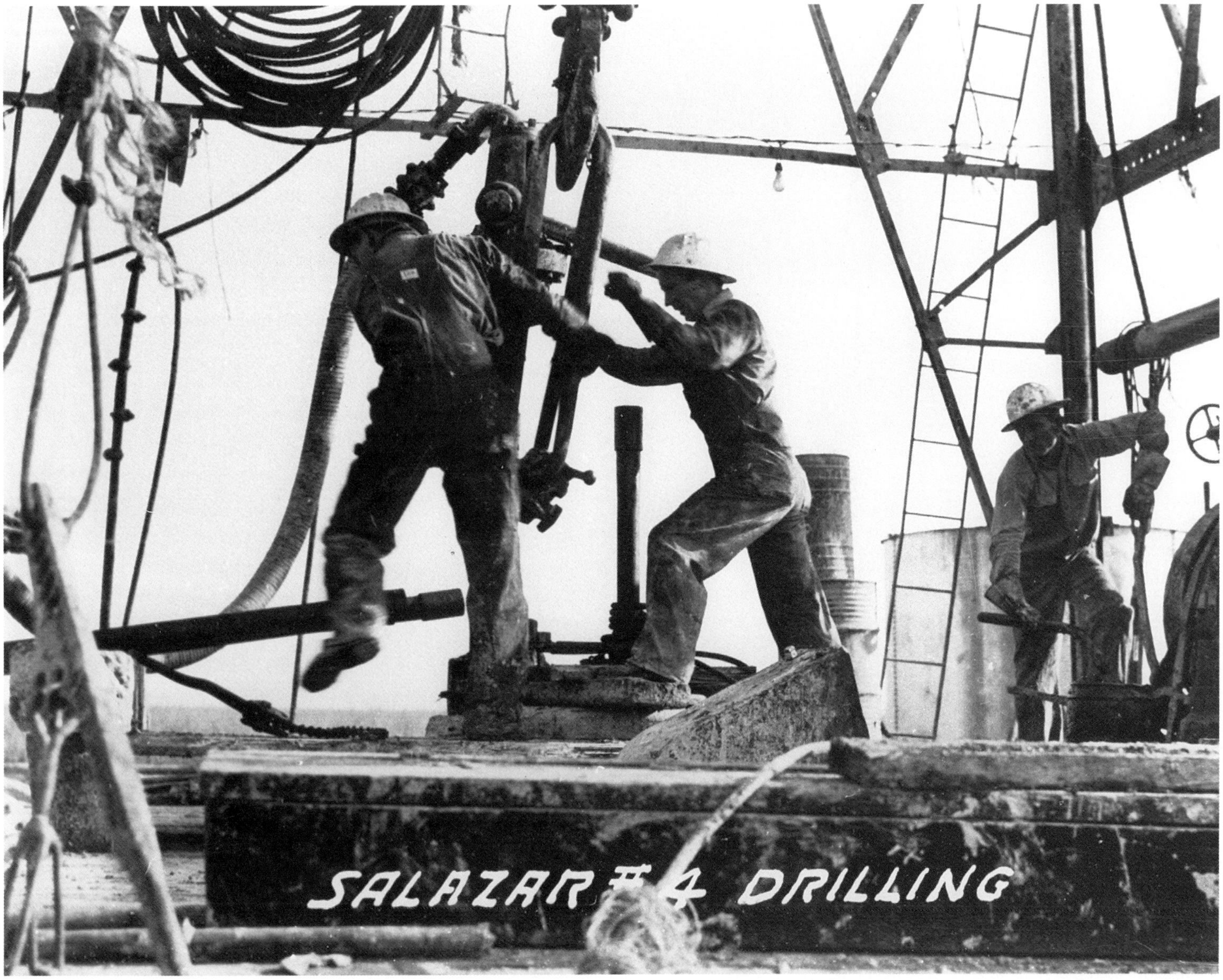

CHAPTER IX

POSTWAR BOOM!

"Even with good production, you couldn't get rich overnight."

World War II changed the lives of many Americans in ways they could never have anticipated. In Texas, the state's economy boomed and expanded in new directions. During the war, new companies moved to Texas after the defense Relocation Act subsidized their shifts from militarily vulnerable locations, mainly on the East Coast. Companies that were later called Bell Helicopter and General Dynamics built large plants and payrolls in Fort Worth to meet the needs of the Army and Air Corps. The Navy and Merchant Marine let large contracts to new shipyards near Houston. The Army PX operation, the largest-volume retailer in America at the time, opened a headquarters in Dallas, enhancing the city's role in distribution of goods. Existing companies, among them Geophysical Services, Inc., and Cameron Ironworks, built specialized equipment for the military, diversifying beyond their business with oil companies and giving Texas an increasingly large foot-in-the-door of

A crew working pipe at Salazar #4 in

Duval County.

A steel rotary rig in South Texas in the 1950s.

the electronics and control systems industries thereafter. Petrochemical plants continued to turn out synthetic rubber and went on to create a dazzling array of new processes and products birthing the Plastic Age.

Surviving veterans bought new houses and earned college degrees with the generous support of the G.I. Bill. Some returned home from the Navy and Air Corps with advanced technical training that fitted them for work in the emerging commercial aviation and electronics industries. Others tried to pick up where they had left off, returning to pre-war jobs and young families.

The oil industry was poised for a new boom when the war ended. Demand for gasoline was up because, with the war behind them, Americans resumed their romance with the automobile, and new and used car sales soared. Trucks began to take longer haul cargoes away from the declining railroad systems as highways improved. If the industry had been able to respond to expanding markets, it would have boomed. The obstacle to a quick postwar take-off was the battery of war-time regulations and controls that lingered after the last shot was fired.

The Petroleum Administration for War began to disband in mid-1945, but it lingered for another year. Worse yet, the Office of Price Administration stayed on, intent on avoiding the stiff postwar inflation predicted by economists, as buying power exceeded the supply of goods available for purchase. During the war, the OPA had permitted the costs of materials essential to the oil industry to rise, but froze the prices of crude oil at ninety-two cents per barrel. Thereafter, it did much the same, when postwar construction made the steel industry profitable; pipe and casing became ever more costly. There was no catch-up for the oil industry until mid-1946, when regulators permitted a long-overdue increase to $1.37. Too little and too late, that amount did little to offset the cost-price squeeze that continued to discourage exploration. Only when a serious shortage of heating oil loomed in 1947 did the regulators permit prices to climb gradually, from $1.62 to $2.32 at year's end. That did it.

The postwar boom began belatedly, but once it began it swept across Texas as oilmen drilled more exploratory wells than they had ever done, reaching to areas that had been considered marginal and into more challenging offshore environments. In the Permian Basin, construction of additional pipelines, capable of carrying the vast resources of the region to refineries on the Gulf Coast, boosted exploration. The discovery of sweeter crude oil at deeper depths, begun in the pioneer Big Lake Field during the late 1920s, continued when two independent companies, Plymouth Oil and Slick and Urschel, drilled to twelve thousand feet in 1947 and found more sweet crude. Confirming the earlier drilling, this Benedum

Field discovery was heralded by Railroad Commissioner Ernest Thompson as the Texas discovery of the year.

Though deeper drilling raised estimates of possible reserves in Texas, it also posed serious problems, particularly for independent oil men. The major difficulty was that deep wells were very expensive, running as high as $300,000 for a single 12,000-foot test. Though the deep pockets of big companies could support costly searches for deep reserves, few independents could raise enough money for deep-horizon wildcatting. What they could do, and would continue to do thereafter, was look for smaller and cheaper projects, the kind that—even if successful— would not add significantly to the reserves of big companies and, for that reason, were left to independents to explore.

From the postwar period to the present, most independents have been "niche players." Sometimes, the scraps have been large. After a subsidiary of Chevron discovered the first in a series of finds on vast underground reefs, for example, independents flooded into Scurry County to lease up farms and ranches before oil scouts and landmen for major companies could tie them up. From Midland, Joseph I. O'Neill, Jr., John Castleman, and William D. Kennedy, among others, raced to lease up uncommitted tracts. W. A. Moncrief, one of the Fort Worth wildcatters who had jumped into the East Texas Field early, did the same thing in the newly opened area. Two years after the discovery, oilmen had completed nearly 1,700 wells and produced 3.5 million barrels of sweet crude per month. There were still opportunities for those who moved quickly.

There were also new cooperative programs that extended opportunity. Early in its productive history the Kelly-Snyder Reef Field began to operate as the Scurry Area Canyon Reef Operators Committee (SACROC) to maintain reservoir pressure and secure maximum yield from the formation. Thereafter, companies increased reservoir pressure with water floods. In 1971 they began large-scale injections of carbon dioxide to retrieve even more of the abundant crude oil in the field. In 1979 the area covered by SACROC yielded its billionth barrel and continued to produce millions of barrels per year thereafter. The SACROC project, undertaken early in the life of the oil field, demonstrated the feasibility of careful reservoir engineering and cleared the way for similar projects in Permian Basin oil fields in years to come.

The Canyon Reef discovery and a backlog of geological data encouraged additional drilling in the Permian Basin with spectacular results. Though Canyon Reef would remain the biggest find of the postwar period, there were scores of additional successes during every year of the late 1940s and early 1950s. One giant discovery

The busy courthouse in Snyder during the reef play boom.

COURTESY OF THE PERMIAN BASIN PETROLEUM MUSEUM.

followed another in the Permian Basin-Plains area, beginning with Levelland, Midland Farms, and Andector in 1945 and 1946, continuing with the Canyon Reef discoveries and with Cogdell and Pegasus in 1949. The region boomed, and for good reason: the success rate for wildcat wells was high, twenty-six percent in 1948, twice as good as the nationwide average. By 1950, the region produced a million barrels of oil per day, comparable to that of Venezuela and second only to that of the United States as a whole. Geographically remote, geologically complex, and previously limited by inadequate pipeline capacity, the Basin came into its own during the postwar boom.

The region also attracted new oilmen who added to the independent exploration community built by Arthur Yeager, Fred Fuhrman, Fred Turner, Bill Penn, George Abell, and others. The newcomers included a group known locally by a generic description, "the Ivy Leaguers." Not all of the oilmen who were described with that term had attended Eastern universities, but they were all new to the oil patch and to the Permian Basin. Men like John and Hugh Liedtke, Earl Craig, Arden Grover, and George H. W. Bush represented a new generation in the Texas oil industry. Some were better connected and wealthier than others. Bush, for one, had to learn the industry from one dusty road and isolated well site to another, starting as a salesman in Odessa, living modestly even when he "went independent" and moved to Midland. The Liedkte brothers, both adept at finance, also settled in and made their fortunes. What the Ivy-Leaguers and other new independents had in common was the "can-do" attitude of veterans and the optimism of wildcatters. With these traits, they began what proved to be spectacular careers. As members of that circle brought in capital from outside Texas and the oil industry, they undertook additional ventures, typically the niche-plays that were still open to independents in a region that the major companies worked extensively. Within Midland, where most of them settled, they were the spark plugs of new community programs—PTAs, the YMCA, Little League baseball, new and expanded churches, and the Republican Party. Many of the newcomers remained through subsequent booms and busts; others moved on.

Oil workers fighting mud on Avenue R in Snyder.

Ragtown in Snyder, 1949.
COURTESY OF THE DELBERT HIRST COLLECTION, SCURRY COUNTY MUSEUM.

In 1953, the Liedkte brothers and George Bush formed the Zapata Petroleum Company in Houston, with Zapata Offshore following the next year. In 1959, Bush split off with the offshore drilling company and the brothers went on to acquire the South Penn Oil Company and its valuable Pennzoil line of motor oils. Their acquisitions continued with the vast United Gas Corporation in 1965 and into additional investments in offshore exploration with the organization of Pennzoil Offshore Gas Operators in 1970. During the 1980s, they won a $3-billion judgment against Texaco as a result of the older company's illegal acquisition of Getty Oil, used most of the gain to purchase Chevron stock, and traded those shares for a huge package of Chevron's domestic oil and gas properties. As a group, the Ivy Leaguers were successful, but the Liedtke brothers were often the most visible winners.

At the same time that newcomers made their fortunes in Texas oil, pioneers in early fields did the same. After the war, Clint Murchison, of Dallas, continued to expand his operations. He had already acquired a small business empire with the creation of the Southern Union Gas Company in 1929 and the American Liberty Oil Company in the East Texas Field. He went on to create Delhi Oil, one of the largest independent integrated companies in the nation, acquired vast natural gas reserves in Canada and merged with Taylor Oil and Gas, extending his investments into insurance, banking, and manufacturing during the 1950s. Along with H. L. Hunt, Algur Meadows, E. Wilson Germany, and other successful oilmen, he helped make Dallas a major oil and gas management center.

From the end of the war to 1960, there were no discoveries elsewhere in Texas to compare with the giant fields that had been coming in throughout the Permian Basin since the mid-1920s, but independents in North Texas continued to deliver up a long list of small discoveries, most of them with reserves of five million barrels of oil or less. The same was true in southwest Texas, where wildcatters continued to drill up the Frio sands and the Vicksburg formation, passing the twenty-million barrel mark only with Mustang Island and Fulton Beach. Some small finds, however, later turned into more significant properties, especially the highly complex Seeligson Field, which had all of the intricacy of a Chinese box, gradually yielding its secrets to patient study. The upper Gulf Coast was relatively quiet as far as oil was concerned, with no discoveries to match those of earlier decades or of the postwar Permian Basin. In the Panhandle, independents continued to work new areas, extending modest production into Ochiltree County with three small finds.

Though the postwar period produced a strong revival in exploration in parts of Texas, it also confronted oilmen with growing challenges, most of them economic. Simply put, by 1956, it cost $3.16 for American producers to replace a barrel of oil in their inventories. If they failed do so, they went out of business just a bit more with every barrel they produced. Their situation was no different from that of a retailer who sold inventory but did not replace it. The problem for oilmen was that the average price they received for a barrel of oil was about forty-cents below replacement cost. Moreover, although important economies had been realized through advanced drilling and production technologies, it still cost oil men seventy cents to produce a barrel of that oil once they found it. That figure was almost four-times the production cost per barrel of oil from the burgeoning new fields of Saudi Arabia, whose wells would yield crude for twenty-to-thirty years at low cost.

Added to the problem of high costs, restriction of production by the TRC further cut into the returns on investment in exploration and production. For example, Midland oilman Fred Turner, Jr., owned wells in the new Noelke (Queen) Field southeast of town. His wells were capable of producing forty-three hundred barrels of oil per day. Proration per well and mandatory shut-down days meant that he was permitted to produce only thirteen hundred barrels per month, about one percent of potential. What these restrictions meant in dollars and cents was that properties that would have yielded $1.6 million per year actually gave Turner less than $16,000. As one Midland oilman put it wryly, "Even with good production…you wouldn't get rich overnight."

Then there were the bad guesses and unwelcome surprises. In the Basin, the classic instance was the discovery and opening of the Spraberry Field. In quick order, Humble, Seaboard Oil, and independents, including Arthur "Tex" Harvey, Harry B. Lake, Ted Weiner, and H. L. Hunt, drilled producing wells into the Spraberry Trend over a five-county region. Another rush was on. The new oil play had a lot of appeal for independents, mainly because the wells were relatively shallow, around seven thousand feet, and they drilled few dry holes. Wells were completed at one-third the cost of those in the Canyon Reef Field, and even at the federally frozen price of $2.58 in 1951, they would pay for themselves in a year or less. Moreover, as drillers and geologists learned more about the Trend, it was clear that it would produce from multiple horizons, another bonus. Early estimates established the size of the field as equal in size to the state of Connecticut and forecast the recovery of more than ten billion barrels of sweet crude. It would be more than four times the size and have five times the capacity of East Texas! It was almost too good to be true. And it was.

As the field developed, oilmen learned that production from Spraberry wells dropped off sharply, even more so than from the old Gulf Coast salt dome wells. Their production forecasts had been much too optimistic. Harry B. Lake, for example, was stunned when production on his four leases fell by more than eighty percent in one year. After his wells were shut in for eight months during 1953, they produced only half that much oil. Two years later, Lake's wells were more costly to operate than their oil was worth, so he plugged and abandoned them. Once expected to provide vast riches, the wells often never paid for themselves. Declining Spraberry wells could be stimulated with acid injections, but even then the results were short lived and it required cash to pay for the process. Three years after its discovery, the geologically complex trend had turned from America's latest oil bonanza, to the country's "largest unrecoverable oil reserve." Thereafter, well-spacings increased and improvements in horizontal drilling and production technologies sustained activity, but it became the arena in which petroleum engineers and property managers displaced explorationists. "Dad" Joiner would have been mystified by this change—now the explorer was less important than the man with a slide rule.

Though black gold was harder to find, natural gas came into its own during the postwar period. Growing production fed Texas' gigantic petrochemical works and

companies discovered it by design, not as an accident on their way to crude oil production. In North Texas, H. Merlyn Christi joined George and Johnny Mitchell and drilled the vast Boonsville Field in Wise County. Thereafter, their company, ultimately named Mitchell Energy, found additional gas reserves in North and East Texas and invested heavily in gas processing operations and connection lines. By the end of the 1950s, the firm was drilling nearly two hundred wells per year. It also diversified into real estate, first in the Galveston area and later by creating the Woodlands, a twenty-five-thousand-acre planned community north of Houston. There was real money to be made in natural gas!

Unfortunately, during the 1950s and 1960s, gas prices were so low that when it was produced along with oil, it made more sense to flare the gas than to invest in gas production and processing equipment, along with the pipelines required to gather it. The dollars-and-cents consideration had always bewildered conservationists. From the time of World War I, some of them had advocated federal control of natural gas production to avoid this waste. State officials were slower to react to it. The Texas Railroad Commission's campaign against the wasteful flaring of natural gas began in Southwest Texas and reached the Spraberry

formation, where two hundred million cubic feet were burned every day. Then, on April Fool's Day of 1953, the commission shut in all wells in the field. The comprehensive closure was not really necessary because a quarter of the oil wells were already connected to natural gas gathering lines so no waste was occurring. Still, to keep the connected wells from producing their neighbor's oil and gas, initially, the regulators clamped down the lid on the whole field.

With lightening speed, oilmen mounted legal challenges to the action, winning a favorable verdict from the Texas Supreme Court only five months after the close-in. Thereafter, the commissions restricted production in the field to ten days a month and required gas connections if significant amounts were burned. That order produced a rush for connections and bargains for natural gas purchasers. Though the TRC's ban was short-lived, proration and declining production led small producers to sell properties in the Trend. The consolidation of ownership improved the economics of producing Trend oil because management of larger holdings tended to be more cost efficient. So, the troublesome Trend lived on to be reworked and traded during coming decades.

Though independents tended to grab headlines for discoveries during the 1950s, the

A crew dressing a bit on a rig. Modern work still requires brains and muscles.
COURTESY OF THE PERMIAN BASIN PETROLEUM MUSEUM.

truth of it was that in the Permian Basin, science and capital gave the large corporations special advantages. They had always mounted the most extensive leasing campaigns in the vast region, which they continued to do. Companies like Humble, Gulf, and Amoco could afford the overhead costs of geoscience and if they found oil, they could wait for long-delayed pipeline connections, which they often provided for themselves. They enjoyed the same advantages throughout Texas, amassing production through exploration, but more often by purchasing leases from cash-starved independents. With these advantages, the half dozen or so biggest producers accounted for about two-thirds of Texas crude by 1955, with Humble leading the list with more than 322,000 barrels per day; Amoco, the runner-up produced half as much.

By this time, there were still also relatively large independents, mainly located in Houston, Dallas, Midland, and Fort Worth. In Houston, Hugh Roy Cullen, John Mecom, R. E. Smith, George and Herman Brown, and Glenn McCarthy controlled companies that sold nearly forty thousand barrels per day. Their Dallas counterparts, Algur Meadows, H. L. Hunt, Ed Cox, and Jake L. Hamon, Jr., were good for thirty thousand barrels. The Fort Worth magnates, including Sid W. Richardson and his nephew, Perry Bass, Kay Kimbell, Arch Rowan, and W. A. Moncrief, held leases that yielded nearly twenty million barrels daily. The total production of the leading independents in the three cities surpassed that of the Sun Oil Company, number six among the top 100 producers of Texas crude, coming in just behind Texaco on the list. These large independents were well known in oil and gas circles, but only Glenn McCarthy, who married into the oil-rich Lee family of Houston, waged a campaign for public notice. He hired Jim Clark, the veteran oil and gas reporter of Houston, as his public relations advisor and succeeded in making both the oil and the society pages on a regular basis. His best publicized venture was actually the opening of the ultraposh Shamrock Hotel in Houston, an event attended by movie stars and royalty, described in living color by national newspapers and magazines. More quietly, John Mecom bought and revived the Warwick, the grande dame of Houston hotels, and Hugh Roy Cullen kept up a steady stream of multimillion donations to nonprofit organizations in Houston.

If Glenn McCarthy's enjoyment of celebrity status led him to seek the spotlight, he was not even a serious contender against another successful Houston oilman, Jim West. West's colorful life style was worth a four-page spread in *Collier's*. He was perfect copy, "...the embodiment of the legend of the big-hatted Texas millionaire." Standing five feet nine inches tall, West weighed 210 pounds and posed in his usual garb, tweed trousers held up by a wide belt, clinched by a four-inch solid gold buckle, a colored shirt decorated by a hand-painted tie that was way beyond "loud," a diamond-set Texas Ranger's star, and all of it topped off by a fifteen-gallon Stetson. He was larger than life, as he smoked a large cigar and held an automatic handgun which he pointed at the *Collier's* photographer. Everything about him was outsized, from the 36-telephone communication system at his River Oaks mansion, to his fleet of 30 cars and three airplanes. Rich men often have expensive hobbies, but West's was truly unusual because his favorite pastime was driving Houston Police Department detectives around in one of his eleven Cadillacs during the detectives' night shifts. He was also fond of tossing hands full of silver dollars, to night club revelers, parking attendants, and, in one instance, to break up a union picket line. As *Collier's* reporter James Aswell noted, Jim West was obviously a square peg in Houston, raising eyebrows when he sauntered into the posh Cork Club at McCarthy's hotel. Still, he was good copy for visiting journalists, and for many Americans Jim West was the enduring image of the "Texas Oilman."

By the time the article appeared, Texas was nearing the end of its expansive oil age. The giant fields had been discovered. The vastest Texas oil fortunes had already been made. The headlines heralded vast discoveries of natural gas, and as Dustin Hoffman was told in *The Graduate*, the future was plastics. Oilmen did well to hang on during the 1960s and 1970s.

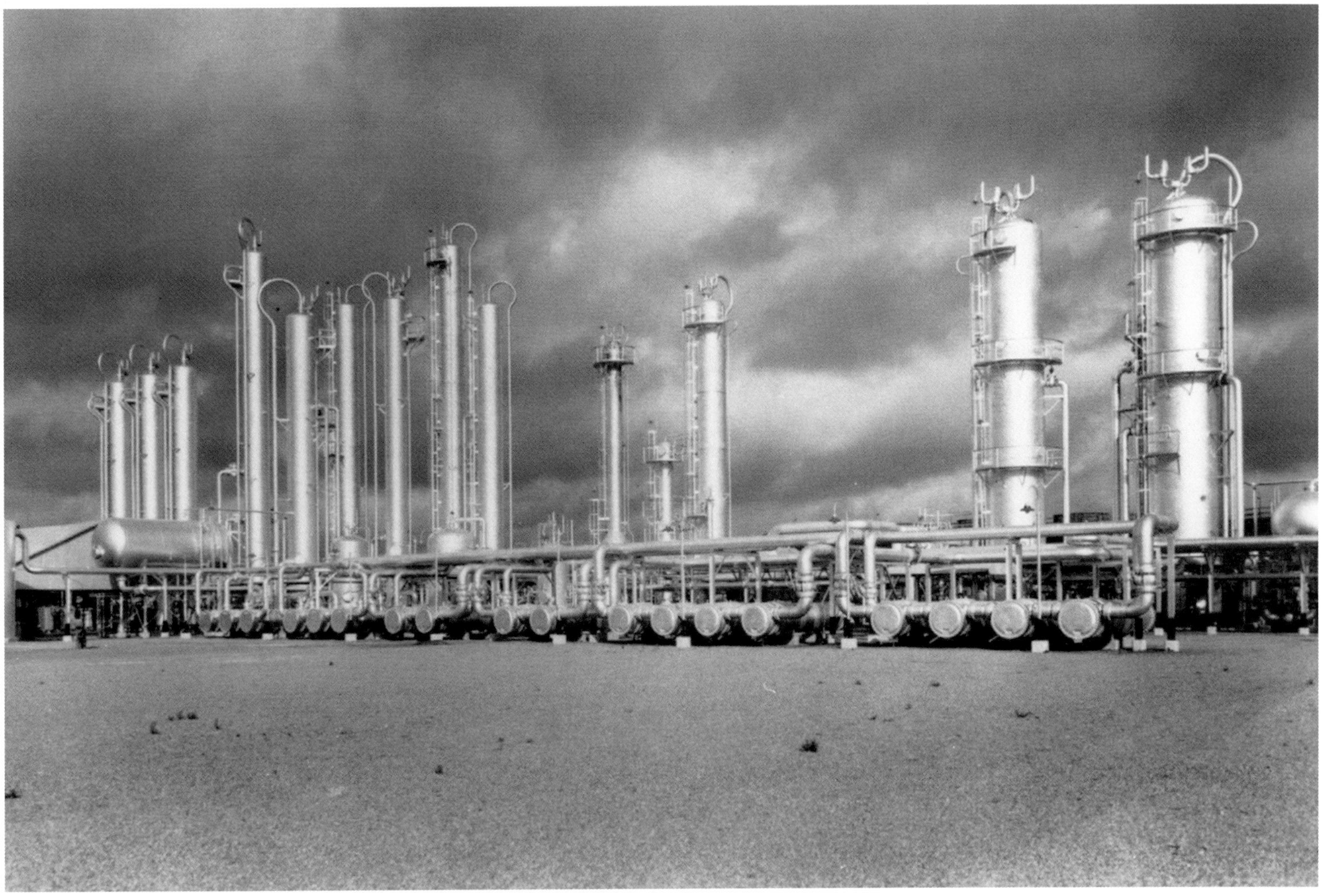

CHAPTER X

MEETING NEW CHALLENGES

"As in Nature, the principle of the survival of the fittest will prevail."
- Michel T. Halbouty on the reality of boom and bust in the oil industry.

The 1960s and early 1970s were especially challenging times, "upstream," for Texas oilmen. The costs of the goods and services they needed to drill wells and operate producing leases continued to rise while the prices they received for oil and gas lagged far behind. Imported oil supplied an increasing part of America's energy needs as the Texas Railroad Commission continued to restrict oil production. The Federal Power Commission also continued to depress prices for natural gas shipped interstate and actually extended its jurisdiction over Texas producers. Under these conditions, the most profitable parts of the industry were "downstream," mainly in refining and petrochemical production, both of which continued to expand to take advantage of cheap oil and gas.

Oil had always been a big-dollar business. For the whole of the twentieth century, one could have provided the capital for a restaurant, a shoe store, almost any small business, for the price of drilling a single oil well. As oilmen drilled ever deeper in their search for new reserves, costs soared. By the late 1960s, it cost three times as much to drill to 12,000 feet as it did to 5,000 feet in most Texas oil fields. When a producer was fortunate enough to have undeveloped or

Amoco's gas cycling plant in Bee County.

underdeveloped leases in a proven field, he still faced escalating costs, up by nearly two-thirds between 1959 and 1972. During the same period, the price of West Texas intermediate crude oil went up by one dime, and then not until 1969. The causes of the cost-price squeeze were the usual ones: proration cut earnings sharply at the same time that the amount of oil available on world markets had increased even more rapidly than demand. The major difference between this situation and that of the 1930s was that America now imported increasing amounts of cheap foreign crude oil from Latin America and the Middle East, all in the interest of sustaining the traditional—if unwritten—U.S. energy policy of keeping oil and gas as cheap as possible to control the costs of goods to consumers, whose purchases fueled the prosperous years of the 1950s and 1960s.

Natural gas was even more severely regulated than oil because the Federal Power Commission expanded its jurisdiction over interstate prices, from user back to the wellhead, with the support of federal courts. The FPC, headed by three presidential appointees, consistently favored gas consumers, including large industries in the northeast, over producers. Still, as increasingly sophisticated geophysical technologies made it less risky to drill in the Frio, Wilcox, and Yegua formations along the Gulf Coast, immense new fields came in from Beaumont to Corpus Christi to McAllen. In the vast Val Verde and Delaware Basins of West Texas, new wells often delivered gigantic volumes of natural gas. The vast Gomez Field of Pecos County, for example, yielded gas and condensate from 1,400 feet of pay; the reserves per acre were nearly 400 million cubic feet. It is not surprising that even in the face of high costs, exploration in Gomez continued into the 1970s; 112 wells yielded 188 billion cubic feet of natural gas. In some fields, there were numerous producing horizons—eight in the large Coyanosa Gas Field. However, even a bonanza gas well was hard to pay for. In 1965, for example, Jake L. Hamon of Dallas and his partners spent $2.75 million over 579 days to complete a discovery

well beyond 21,000 feet. The development wells drilled to follow up on his discovery cost $1.75 million each. These wells were obviously far beyond the "poor-boy" efforts of earlier decades because they required vast capital and enough money in the bank to wait for a slow return of the investment. Even a successful well might earn less than the same amount of capital would have received if the oilmen had put the same money into United States Savings Bonds. Beyond the familiar risks of dry holes and small wells, gas found at great costs and sold at low regulated prices put an oilman out of business. As one successful producer put it, "You can lose your ——ing shirt in gas."

With some of its wells drilled nearly three miles into the earth, the Permian Basin was high-cost and high-risk for explorationists. The Panhandle and South Texas were less so. During the 1960s, Amoco and Chevron followed Phillips' lead in extending the vast Panhandle Gas Field, bringing in strong production from about seven thousand feet. In South Texas, Slick Oil and Chevron completed big wells, each producing more than 120 million cubic feet of gas per day from formations that were found short of 10,000 feet. Near Laredo, smaller volumes were located at depths of less than three thousand feet in 1968. The lower cost and risk of searching for oil in South Texas encouraged additional exploration to the extent that it was the principal natural gas province in the United States by 1970. Operating out of Houston and Corpus Christi, Texas operators kept rigs running steadily in South Texas after the war.

As risky as gas exploration was, in the absence of sizeable oil prospects, one had the choice of accepting that challenge or going out of business—many oilmen either sold out or were washed out. Around the country, the number of independents dropped from about 42,000 at the beginning of the postwar boom, to fewer than 15,000 by 1972. Some of the smaller independents sold their assets, closed their offices and found new occupations. The larger ones sought and accepted take-over bids. Large and relatively efficient companies, such as Honolulu, Union Texas Natural Gas,

Republic Natural Gas, and Plymouth Oil closed their doors. In all, between 1952 and 1962, of 31 independent producers large enough to be listed on the New York Stock Exchange, 15 were bought up by the survivors, usually large companies such as Mobil, Humble, Marathon, and Monsanto. During the next three years, from 1963 to 1965, there were more than 150 big sell-outs. High costs and tight regulation brought tighter ownership of oil and gas than any group of conspiring monopolists could have accomplished.

Independents did not accept crippling regulations passively. During the postwar period, for example, Texas independents, led by H. J. "Jack" Porter, of Houston, organized the Texas Independent Producers and Royalty Owners Association to push for the end of federal price controls. Thereafter, TIPRO fought for the interests of its members in Washington and Austin, with more success at home than in the nation's capitol. Price controls were lifted in the late 1940s, but TIPRO along with other groups was unable to stem the growing flood of imported oil until late in the Eisenhower administration, when voluntary restraints were enacted into federal law. In Austin, TIPRO was usually successful, fighting unreasonable production limitations by the Railroad Commission, heading off the forced unitization of properties, and discouraging additional taxes on oil and gas production. Across the state, regional groups, such as the Panhandle Producers and Royalty Owners Association, the Permian Basin Petroleum Association, the North Texas Oil and Gas Association, built their memberships and worked for the interests of their members in Washington and Austin. More naturally competitive than cooperative, oilmen organized to survive.

Apart from the trade associations, an increasing number of oilmen took an active interest in politics. Jack Porter, for example, ran as the Republican opponent to Lyndon B. Johnson in the 1948 race for the United States Senate. Four years later, Porter was elected Republican National Committeeman for Texas, taking the party thereafter through a critical phase of development, away from its traditional core of aspirants for federal jobs and sharply in the direction of conservative ideology and competitive politics. Under Porter and his associates, Republicans worked hard to elect Dwight D. Eisenhower in 1952 and 1956, swelling the ranks of active party members with a vigorous new generation of activists. Among them, George and Barbara Bush and other bright and energetic young couples participated in party politics for the first time and worked at winning, often using *How to Win*, a guide to effective electioneering developed for northern Democrats by the United Auto Workers. After he moved to Houston, Bush was elected to two terms in the United States House of Representatives in 1966 and 1968, beginning with what proved to be a long and successful career in national affairs as ambassador, CIA head, and vice president and president of the United States.

In the meantime, oilmen sustained the Republican Party through the 1960s, with new leaders, including Peter O'Donnell, a Dallas investor, and newly elected U.S. Senator John G. Tower. Scattered successes in Dallas, Houston, and West Texas began to add up as the oilmen raised money and recruited candidates while their wives swelled the growing ranks of the Republican women's clubs and did the demanding organizational work of turning out voters for Republican candidates. Rita Bass Clements and other Dallas and Houston Republican women were so effective that they were often tapped for national party positions. As time passed, the Republican women, in particular, became highly skilled at voter canvassing, working telephone banks, and the other time-consuming tasks that were essential to success. By the 1970s, oilmen and their wives had done more than help keep the party alive—they did a good bit to elect one of their own, William Perry Clements, Jr., to the governorship in 1978 and again in 1986. Clements was the first Republican to hold that office in a century.

The next generation of activist oilmen produced more party leaders, such as Ernie Angelo of Midland, and Ray Barnhart, a North Texas legislator. Victories at the polls continued

to mount as George W. Bush, a Midland native and one-time oilman, was elected to two successive terms as governor during the 1990s. By 2002, Republicans had come from their long-time status as the nominal opposition to the Democratic Party in Texas before 1952, to holding control of both houses of the legislature and all of the general offices of the state. Not that independent oilmen and their wives did it all—but they were often at the head of the parade.

Would the oilmen and their wives have become political activists if federal regulations and policies were not of tremendous importance in the industry? Probably so. Wealth, business and social connections, and leisure time to give to volunteer jobs would probably have put them in the ranks of the Republican conservatives. Still, it seems likely that the policies of the Federal Power Commission, the depression of domestic prices by foreign imports, and such tax issues tied specifically to oil and gas as the depletion allowance—a credit for the reduction of assets as oil was produced—made them all highly sensitive to politics at the national level. The combination of the cost-price squeeze and controversial federal policies and regulations threatened the independents, and they fought back through industry groups and at the polls.

They also coped with the inevitable problem of declining production from their oil wells. By 1968, Texas operators owned nearly 90,000 stripper wells, each producing 10 barrels or less of oil per day. Individually, the wells were of minor importance in the total production of the state, averaging about 4.5 barrels per day, but together produced from reservoirs that contained 2.3 billion barrels of recoverable oil. About two-thirds of the wells were pumped— they had long since stopped flowing. The remainder was in a stage of development usually referred to as "secondary recovery." Generally, that process required the injection of either water or natural gas to increase pressure in the formation, thus pushing more oil toward the wellbore, where pumps raised it to the surface.

Oil fields in Pennsylvania had been coaxed along from the 1880s onward. The design of injection programs became more sophisticated over decades, to the point that most were worth the additional investment in equipment and services. Over time, however, a repressurization project yielded ever less crude oil. The typical stripper well in Pennsylvania pumped only a quarter of a barrel per day in 1968. By that time, the planning and calculations that were behind these programs were ordinarily the work of petroleum engineers, trained in the technology and economics of reservoir management. As more and more wells needed enhanced recovery programs, the engineers came into their own in the oil industry and the state's university's—Texas A&M and the University of Texas at Austin early on— created important undergraduate and graduate training programs, adding to their nationally recognized geology departments and research programs. The challenge to their graduate engineers was to design and operate enhanced recovery programs that made money even while production was restricted and prices were low.

Not everyone suffered when oil and gas prices were low. The processors—refiners and petrochemical manufacturers in particular—often did well because the costs of their raw materials, usually called "feedstocks", were essentially frozen while their products increased in value. That was one of the advantages of being a fully "integrated" company; if you lost money as a producer, you gained at least some of it back as a buyer and processor. Moreover, if you were as big as Exxon, Chevron, Texaco, and other firms, you had access to cheaper imported crude oil.

For good reason, the big companies continued to build on new petrochemical plants and to expand their refineries on the upper Gulf Coast. By 1953, about half of the refining capacity of the region was on the Houston Ship Channel. The once-wasted gasses given off as by-products of refining were carried to petrochemical installations for additional processing. During the 1950s and 1960s, the Houston and Beaumont

Midland's most celebrated Ivy Leaguer, George H. W. Bush, confers with his driller in this photograph from the early 1950s.

newspapers carried a steady stream of articles describing the new installations and the massive investments they required. Gulf poured money into its Port Arthur refinery making it the third largest in the nation by 1955. Mobil invested hundreds of millions of dollars upgrading and expanding its works at Beaumont, matched by Exxon's continued revamping and expansions at Baytown. Texaco, Pure, Atlantic, and Sinclair continued to build and to invest in complicated new technologies. By 1955, this sector of the industry employed more than fifty thousand Texans. There were more than sixty refineries in Texas, but more than half of the total capacity was in six largest installations of Humble, Mobil, Gulf, Texaco, Amoco, and Sinclair. Like oil and gas production, refining was increasingly dominated by the largest firms.

In 1955, the petrochemical plants, supplied with cheap crude oil and natural gas through Texas pipelines, continued to expand with seventy-five percent of all new American capacity added in Texas. Goodyear and Goodrich made vast quantities of synthetic rubber at Beaumont. DuPont, Allied Chemical, and Kodak Eastman built plants at Orange. Houston Chemicals erected a tetraethyl plant next to Mobil in Beaumont to use chemicals that flowed from the refinery. The Golden Triangle, Beaumont-Orange, Port Arthur, and Houston-Texas City had 11 refineries and 42 petrochemical plants all connected by a spaghetti bowl of pipelines. Seven chemical plants made ethylene which they piped to 29 other plants through 625 miles of product pipeline.

The inter-connection of the increasingly vast refinery-petrochemical complex can be illustrated by one of the simpler systems, one that originated at the Shell refinery at Houston. From the refinery, Shell moved refinery gases to its chemical plant, where they were converted into propylene and isopropyl alcohol, from which it made acetone, which it moved to Rohm and Haas, which made acrylic monomers. That product moved by barge and tank car to other manufacturers who produced plastics. With this efficient system of producing

and transporting products to other manufacturers, the Gulf coast refiners and petrochemical producers grew rapidly, dominating the U.S. market, and they kept growing. Mobil created an ethylene line at Beaumont that was the largest in the world in the early 1960s—that is until it was surpassed by new Monsanto and Phillips plants during the same decade.

By 1965, there were two hundred petrochemical plants in Texas, all of them developing new processes and finding ways to cut costs. Beginning in the 1960s, refiners and petrochemical producers also found ways to use gases that were once discharged into the air, reducing smoke emissions to conform to new federal environmental regulations and cutting some costs at the same time by using once wasted gasses to generate power. The refiners and petrochemical managers adopted computer technologies during the 1960s as well, adding capacity and cutting costs, often by reducing employment. Shell added new hydrocracking capacity to its Houston refinery in the late 1960s cutting costs and providing additional feedstocks for Shell Chemicals at the same time the other big companies were doing much the same thing. From the 1950s to the 1970s, the downstream processors doubled production, became more efficient manufacturers of gasoline and petrochemicals, and cut their payrolls by one-fifth. There was money to be made in plastics, indeed! But it took money to make money.

By 1967, the petrochemical complex of Texas employed 45,000 workers—down ten percent since 1955—and the works had cost more than $5 billion to construct. The effort was impressive, whether judged by the miles of pipeline, paychecks issued, or by the commanding position the Texas industry had in the production of synthetic rubber—more than eighty percent of American production—and of such basic products as styrene, benzene, and ethylene. By the end of the 1960s, refining and petrochemicals were employing as many Texans as the exploration and production phases of the industry. It was a new day.

CHAPTER XI

"THE LAST BOOM?"

"If there is one lesson that should have been learned by the entire industry over the last couple of years, it is that the art of predicting the future of the oil and gas industry has been found to be woefully lacking."
- Ray L. Hunt, 1985.

Most writers have learned that it is dangerous to identify any event as either "the first" or "the last." Firsts are always challenged because there is often some happening that seems to have come earlier. In the story of Texas oil, for example, the discovery of oil has been pushed back from Nacogdoches during the 1860s to observations by Luis de Moscoso, a survivor of the Desoto expedition, in 1543. Nothing came of his report, but it appears to be a "first." "Lasts" are also risky. James A. Clark and Michel T. Halbouty wrote *The Last Boom*, a lively account of 1930s turmoil in East Texas. The book had barely appeared in print in 1972, before Texas oil entered another frenzied phase and moved from boom to bust during the next fifteen years. Surely, that must have been the last boom!

Unlike earlier oil booms, the most recent one was slow in developing and it was triggered by politics and economics, not by the discovery of vast reserves. There were no new fields like those at Spindletop, Ranger, Borger, and Wink. There were no wildcatters like Anthony Lucas, A. E. Humphreys, or "Dad" Joiner. The new boom was fueled mainly by economic incentives that resulted from dramatic changes in the control of world oil and from energy policies of the federal government. The risks oilmen tried to manage were also different, found less in the traditional high probability of a dry hole than in changes in world markets and government policies over which they had little, if any, control. It was a new age, with dangers that were not nearly as evident as new opportunities. If there is an appropriate sports analogy, the action was fast and continuous, more like soccer than baseball.

Congress kicked the ball the first time, when it reacted to declining U.S. production by raising controlled prices on "old oil" from reservoirs in production before 1973. The increase was significant—from about \$3.15 to \$5.05 per barrel. In itself, that boost would have stirred oil field activity, at least to sustain production through enhanced recovery programs. The Economic Stabilization Act went farther, permitting an average of \$10.82 for new and stripper well oil. At the end of the year, the ceiling on old oil was raised by an additional dollar. Within Texas, the new price incentives, reinforced by the setting of allowable production at one hundred percent, stirred activity from Houston to Wichita Falls and from Amarillo to Midland and Corpus Christi.

Revival of the industry was also triggered by events half way around the world in 1973, when Libyan strongman Colonel Muammar al-Qaddafi, used U.S. support of Israel in the Yom Kippur War to rally fellow Arabs and other members of the Organization of Producing and Exporting Countries (OPEC) to take control of the price of oil produced in their countries away from the large integrated international oil companies—the major buyers.

A trailer court on barren land, such as this one in Midland in 1983, was a familiar sight during the 1980s boom.

When oil prices on world markets soared to $36 a barrel, federal policy makers were faced with two unacceptable alternatives: they could let domestic prices rise to that level, triggering runaway inflation and weakening industries that were heavy consumers or energy or they could put a lid on U.S. crude oil prices and, thereby, discourage exploration and development, making the country even more dependent on uncertain supplies of costly foreign oil. It was a lose-lose situation. With no fault-free option, the federal government pursued both policies at the same time, permitting the price of "old" oil to rise by $1, while "new" oil and crude produced from stripper wells were permitted to rise to the 1973 world-level of about $11 per barrel. In theory, this approach would both limit the effects of higher energy costs and encourage exploration for reserves within the United States. In fact, it worked well for three years, setting off a significant increase in searches for new reserves.

This old oil field housing was hauled into Odessa, where it is still occupied. The three-room shotgun house was lived in during the boom of the 1980s.

Geophysical work increased sharply. Drilling contractors had all the work they could schedule. Labor was in demand and wages rose. The oil industry spent three times as much drilling for new oil in 1976 as it had two years earlier, and that was just the beginning. In the Permian Basin, wildcatting rose by 22 percent in 1974 and by 12 percent the following year. Well completions increased by nearly one-third in 1975. Unfortunately, inflation also soared, reaching double digits

during the presidency of Jimmy Carter. In response, Congress passed the Energy Policy and Conservation Act of 1976 and restructured price controls. Now, stripper well production would continue to fetch world market prices, then above $14 per barrel, but new oil would bring a dollar less while old oil was held to $5.25. The impact on the petroleum industry was immediate. Wildcatting dropped off, well completions declined, and independent oilmen cut back on their searches for new reserves. Near the end of the Carter administration, oil prices were permitted to climb, reaching about two-thirds of the world level.

Activity rebounded and took off after prices were deregulated following the inauguration of Ronald Reagan. By mid-1981, prices reached $31.77, a ten-fold increase—not adjusted for inflation—over what they had been just ten years earlier. During the decade, the Federal Power Commission, alarmed by the increasing flow of gas into unregulated intrastate markets, permitted prices to rise, from 52 cents in 1970 to $1.53 in 1977. The next year, the natural gas policy act deregulated gas produced from deep formations, expensive to drill and operate. Prices for gas from tight sands, through which gas moved slowly, were doubled. And the boom was on.

What the government gave, it also took away—at least in part. The new provisions for gas came with 365 pages of regulations in 1978. An operator who found gas that qualified for higher prices had to shut in his wells until the Federal Energy Regulatory Commission processed his paper work. While he waited for red tape to be untangled, the oilman waited and tried to pay his bills. The increase in prices of oil was also offset by a take-back, a steep tax on the "windfall" profits generated by deregulation, though it was actually levied on income and not on profits. Texas Senator Lloyd Bentsen correctly described the scheme as "one more sorry example of the government offering the carrot of decontrol and then taking it away with the stick of punitive taxation." Though the act was to be phased out at the beginning of 1988, Texas congressman Bill Archer reminded his constituents that "Once you put a tax in place and the government begins to depend on those revenues, it becomes very difficult to change." Archer and Bentsen were on-target.

There were other unpleasant realities that offset the bonanza returns that should have accompanied the boom. At every step, costs soared. The first one, leasing, went out of sight. On University lands, for example, from 1973 to 1980 the bonuses bid per acre increased nearly nine hundred percent. Contracts routinely raised the royalty interest, from one-eighth to one-quarter—with a few landowners in the tight-formation Austin Chalk area of central Texas receiving one-half. The increased demand for services and equipment brought sharp jumps in costs for drilling and workovers. In a single year, 1978, the cost of drilling a deep gas well in either the Permian or Anadarko Basins rose by more than one-third.

Borrowing money, necessary to finance operations at most companies also became vastly more expensive as interest rates edged upward from 12 percent in 1974 to more than 20 percent in 1981. With rates so high, banks and other energy lenders made as many energy loans as they could clear, usually staying within conventionally conservative lending guidelines, loaning half the value of reserves and equipment based on estimates of their value when the loans would be repaid. Just to be on the even-safer side, some banks required that oilmen keep as much as twenty percent of the amount of their loans on deposit at the banks, making it possible for the bankers to re-lend that money for additional profit.

What it all added up to—increased costs, of services, materials, leases, and money—was that they largely offset the higher prices oilmen received. In 1981, putting borrowed money in an expensive well made no wildcatter rich. But, given the steady rise of prices on world markets, it was reasonable for oilmen and their bankers to expect that trend to continue. Most energy economists agreed that prices would hit $80 per barrel oil by the end of the decade and $90 a barrel oil thereafter. When the soaring costs and real risks of drilling were weighed against those projected prices, a well that was expensive to drill in 1980 would pay out handsomely in less than a decade.

Oilmen have tended to be optimists by nature, or they would never try to make their fortunes in that high-risk industry, so they, their investors, and bankers, were not discouraged by sky-rocketing costs. At $90 per barrel oil down the road, all costs were affordable, all loans would be repaid, and most oilmen would roll in cash. At least, that was consensus in the oil patch. As a result, with considerable inducement from federal policy makers, oilmen drilled extremely expensive wells looking for tight sand gas, which required still more money to produce; they drilled prospects that they would never have touched in the 1960s, when prices were low. They searched for reserves with increasingly sophisticated and expensive geophysical processes, drilling for smaller yields of oil and gas than would have attracted them during slower times. In the common language of explorationists, money was chasing deals, a situation that is not likely to produce success in any business. Drilling prospects that had been rejected as "dogs" were dusted off and contracted.

Areas that had not panned out in earlier booms were drilled again. The Austin Chalk formation of central Texas, for one, had not been tested since the 1920s, but the wells were either small or they played out quickly. Higher prices and new technology brought that play alive. Wells in the tight limestone formation were completed more rapidly because of improvements in drilling technology and equipment. Then, using hydrofracturing and other techniques, operators boosted production. Even with this treatment, however, the wells often declined by 50 percent within six months. In the meantime, they produced 50,000 to 80,000 barrels of oil, except in the "sweet spots," in which independents such as Clayton Williams, Jr., of Midland had better wells. Williams did even better in the difficult field because he completed a gathering system during the early life of production and profited from this second source of revenue. Other independents brought in non-oil investors with "can't miss" promises, with the assurance that they would be correct in the literal sense that there would be oil in their wells. What they were less sure of was that they would produce sufficient oil to even recoup invested capital, let alone show a profit. Booming Giddings, the center for service companies in the Austin Chalk, was overrun by men talking and breathing oil. Traffic mounted to the extent that additional stop lights were placed on its main street, making it almost as time-consuming to drive through as Austin.

The Spraberry Trend in the Permian Basin revived, in part, because its wells nearly always produced oil, though commercial production was sometimes short lived. With consolidation of properties, operators such as Midland's Parker & Parsley, Inc., achieved economies of scale. The application of water flood and water injection, of as much state-of-the-art engineering as wells could pay for, also coaxed still more oil out of the ground. By 1981, the problem field had produced more than six hundred million barrels of oil, going on to reach one billion barrels in less than twenty additional years. There were good reasons for oilmen to be happy that they had stuck with the Trend.

Optimism was the mood of the day. However, not everyone in the Texas oil and gas community shared the general enthusiasm. One independent, owner of large natural gas reserves, described the boom as "tulipomania," like the seventeenth century craze during which Dutchmen bid up prices for rare tulip bulbs at a dizzyingly and self-destructive rate. For some independents and workers, remembrance of times past prompted increasing caution. John J. Redfern, Jr., a successful Midland independent who started in the industry during the Depression, paid off his loans and cut back on exploration as costs soared. In nearby Monahans, a thirty-year-old tool pusher was pulling down better than $90,000 a year, lived in a double-wide mobile home, and saved much of his income because his father, another survivor of the Depression, warned him that the good times were not going to last.

That was a minority opinion at the time. Most businessmen in oil-producing regions agreed with Vice President Bush's sons, George W. and Neil, who entered the industry in Midland and Denver, respectively. George jumped into the business in 1975 with Arbusto Energy ("arbusto" is "bush" in Spanish.) At the height of the boom, in 1982, he created Bush Exploration, which merged with Spectrum 7 as prices declined in 1984. Thereafter, he sold out to Harken Energy, a Dallas firm, and decided to take his chances on bum calls by baseball umpires when he became CEO of the Texas Rangers. After two successful campaigns for the Texas governorship, he moved along to Washington, D.C., as president. Bush never made a fortune in oil, largely because he entered the industry too late to acquire the cheap reserves that provided steady income for industry veterans; every well he drilled cost top-dollar during the boom. When oil sold for more than $30, Bush's investors made money, but return of their capital, let alone profit, depended on the continuation of high prices and future rises.

The tool pusher's father was right, but fathers do not always know best, and some of the most successful oilmen in Midland continued to expand their operations—and risks—into the 1980s. Money flowed—into oil deals, real estate investments, palatial homes, personal jets, and expensive automobiles. Newspapers and glossy magazines frothed with descriptions of the new rich. Midland's Rolls Royce agency was good for mention whenever reporters from Dallas, Houston, New York, and Washington, D.C. wanted a startling example of new riches. With less than one hundred thousand people, Midland had a Rolls agency! Critics of the oil industry, always a crowd on the East and West Coasts, likened the exuberant oil rich to Belshazzar, king of ancient Babylon, whose lavish feast was interrupted when a ghostly finger etched a warning on a wall: "You have been weighed in the balances and found wanting." Wanton consumption invited divine retribution.

Godly wrath was not needed to end the boom—changes in international markets did the trick. Major recessions in the early 1980s curtailed demand, which had also dropped in response to consumer and industrial cuts in energy use. Consumers lowered their thermostats at during winters and raised them in the summers. Energy-efficient construction and manufacturing cut energy needs as did production of energy-efficient automobiles—for a while. By the end of 1982, it was clear that the boom was over, at least for the time being. Drilling declined, lay-offs of both blue and white-collar workers accelerated, and many of them hit the road to search for new jobs. Midland and Odessa, between them, lost more than fifteen thousand jobs during the decline. Bankruptcies climbed by twenty-five percent.

Though there was a strong sense that the days of wide-open expansion and quick fortunes were over, most economists believed that the market changes reflected price adjustments and that oil would now be sold for prices between $25 and $35 per barrel. That spread, of course, was so wide that the $10 difference was often more than enough to determine if a company stayed in business and geologists and drillers kept their jobs. Those who survived hung on for better times.

They didn't come. In 1986, the bottom dropped out of the international crude oil market, sending prices to $12 and lower. The impacts on the industry and on Texans were widespread and devastating. The story was repeated many times and in many different versions. In Odessa, a forty-year-old worker lost his job as a tool pusher in 1982 and became a driller for less money. One year later, he lost that job and began to work as a roughneck. That work disappeared in 1984, so he took on occasional unskilled jobs as a roustabout. Finally, in 1986, those jobs dried up too and he was jobless, sharing his mobile home with an unemployed son, other children, and grandchildren. In Corpus Christi, a man nearing fifty years of age lost his rig supply business in bankruptcy, took odd jobs until he was hired on at a 7-Eleven convenience store, then lost that job when the store closed. Exploration declined sharply, leaving a quarter of the geologists in Texas unemployed by the end of the year. An oil field bumper sticker said it all: Please, Lord, let there be another boom. I promise not to piss it away!" As catchy as it is, the slogan got the story wrong. Oilmen who lost their companies and workers who lost their jobs did not fail because they had lived too well when times were good. Had they spent money more carefully, perhaps more of them would have held on longer, but even then, volatile oil prices would have wiped most of them out in a few more years, at best. It was a feast-or-famine industry and the seven fat years were followed by more than seven lean ones.

Ripples of the downturn were obvious in most parts of Texas and they spread to other businesses. In Houston, Dallas, and Austin vacancy rates in office buildings climbed to twenty-five percent and beyond while new home construction ground to a halt. Banks stuck with large energy and commercial real estate loans faltered and failed, beginning with the banks in Abilene and Midland in 1983, and including all of the mega-banks of Dallas and Houston. Large out-of-state holding companies took over their operations, leaving only Frost National Bank of San Antonio, small by comparison to First National of Dallas or Texas Commerce of Houston, surviving as a substantial Texan-owned bank.

This modern cot house served workers in Midland in 1983.

Governments and other public agencies also took hard hits. Severance taxes on oil and gas at the wellhead plummeted, with sales tax revenues declining along with the fall off of drilling and service business. The State of Texas lost more than $1 billion in tax revenue because of declining oil prices and rising unemployment. Budgets and programs were cut and public employees were laid off, joining the scientists, drillers, and roustabouts at the Texas Employment Commission in searches for temporary jobs.

Oilmen sought remedies where they could find them, selling properties at hardship prices to pay at least some bills. The most optimistic Texans even expected the federal government to provide some relief. Energy Secretary Donald Hodel explained that the Reagan administration was sympathetic, but that further study of the security implications of the energy disaster would have to be completed before either Congress or the executive branch would propose solutions. "Further study" was cold comfort! In the oil patch, a parade of well-paid motivational speakers peddled upbeat perspectives, suggesting that oil company employees who had lost their jobs, or expected to, and wildcatters who had failed or were about to, listen to comical audio tapes, sing in their cars, and be more spontaneous. Feel-good psychology was no more useful than sympathy from Washington. These attempts to be upbeat in the face of financial disaster were reinforced in Midland by a pep rally, with bands, cheerleaders, and the release of three thousand helium balloons. The balloons rose, but oil prices did not. Whatever the beneficial effect of positive thinking, it was short-lived. Even the irrepressible humor of Texans took a grim turn: "What's the difference between an oilman and a pigeon? A pigeon can still make a deposit on a Mercedes."

By the end of 1986, it was clear that there would be no dramatic reversal of the oil industry's fortunes. Sales of drill pipe, rigs, and other goods and services had fallen from $40 billion in 1982 to $9 billion. Employment in the upstream sector of the industry fell from 100,000 to 25,000. Local economies were hard-hit. By 1986, there were two hundred thousand vacant homes in Houston, twice the normal number. Repossessed drilling rigs, seized by failing banks sold for ten cents on the dollar—when there were buyers.

It took Texas nearly eight years to recover from the devastating bust that began in 1982. Not until February 1990 did employment and other indexes of economic activity return to the high point reached in March 1982. Even then, life was not the same. Jobs lost when troubled companies merged or acquired were never recreated. The once vibrant oil and gas community would adjust to the hard realities of a diminished industry and to the fact that the directions it would take in the future often would have little to do with hard work and new ideas.

The Golden Triangle region of Beaumont-Port Arthur, Orange, and Baytown-Houston also lost jobs. Dresser-Ideco, a Beaumont drilling manufacturer cut its payroll from 1,000 down to 350 workers. Workers at oil-related construction companies took twenty-five percent pay cuts, when they kept their jobs. Texaco laid off fourteen hundred workers at its Port Arthur refinery. In union hiring halls, members played cards and dominos, with scant hope of being rehired. As profit margins at refineries and petrochemical works weakened and then disappeared, those companies laid off employees and introduced labor-saving technologies to regain profitability. In mid-1986, only twenty-five percent of the active members of the Labor International Union in Port Arthur were fully employed. Labor unions in the area had weak bargaining positions and generally accepted scale-downs and automation without striking. As a representative of Gulf refining put it, "...unions here, to their credit, are flexible. And they're learning to be more flexible...." As journalist Louis Bose described Port Arthur in 1986, the only growing operation was a local soup kitchen, which served about 250 hot meals a day.

In East Texas, Lone Star Steel laid off three thousand workers when demand for steel pipe fell with the collapse of drilling activity. Unemployment in the Longview-Marshall area reached about eleven percent. Unemployed scientists in the Permian Basin and Houston went back to school to earn teaching certificates, training for more stable positions that would often pay substantially less than half of their onetime oil-industry salaries. One highly skilled offshore drilling supervisor lost his job when a Houston-based company cut back sharply. He found another job with the United States Postal Service, and kept it even when his telephone rang again during later upturns in oil. For the most part, the oil industry lost these men and women and their valuable experience permanently.

There is no honest way to put a happy face on the bust. It was the ill wind that blew no good. It was a wrenching experience for displaced workers and their families and it was disquieting for the survivors, many of whom worked with less confidence in the security of their own jobs. For the state as a whole, only the rapid rise of the high-tech sector of its economy offset the disastrous losses that spread across the state when the boom was over. Even then, in future decades, growth in that industry would prove to be cyclical, leaving additional unemployed workers and managers in the wake of its own bust. The oil industry would experience subsequent revivals, but it never regained the place it occupied in Texas life before 1986.

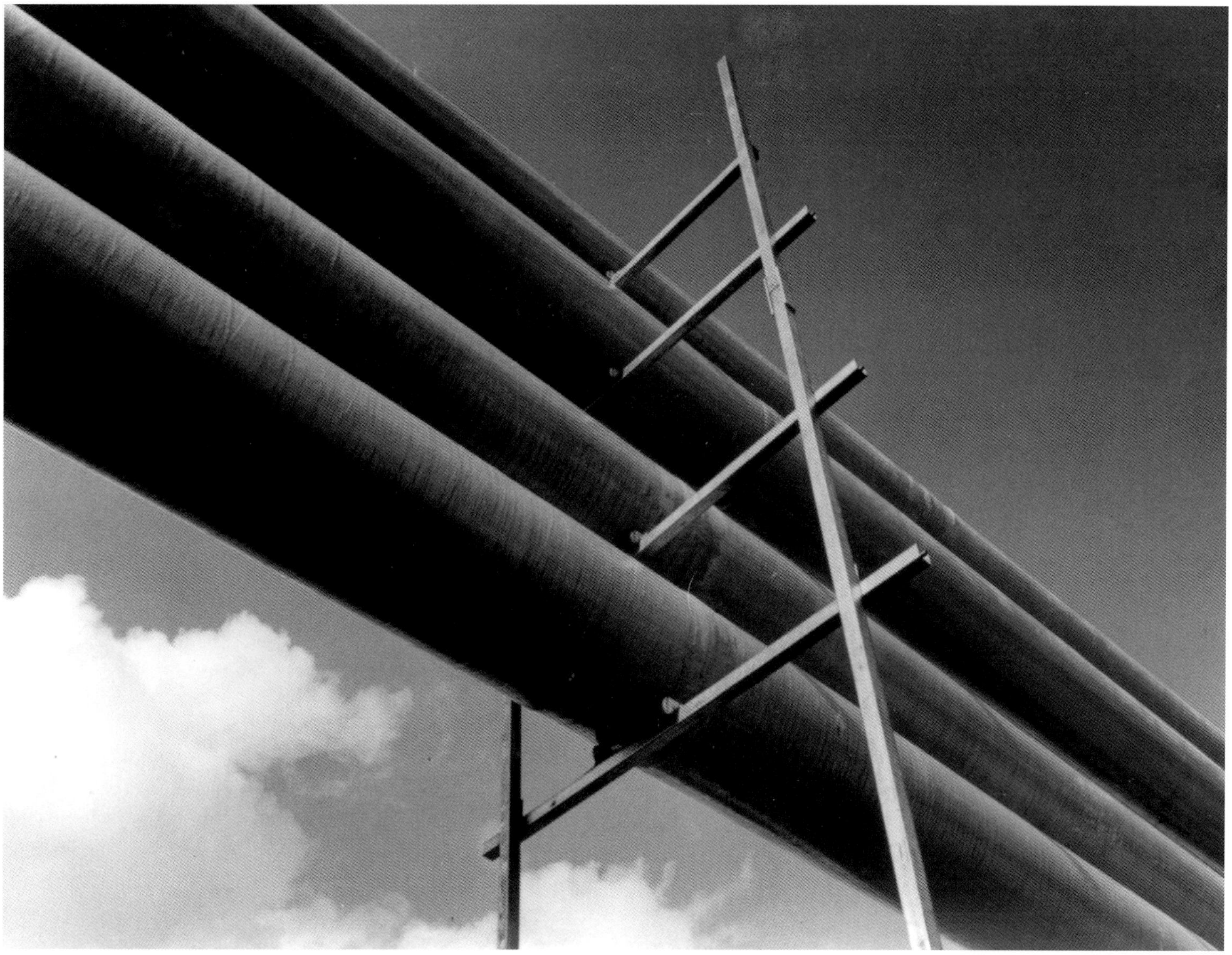

CHAPTER XII

TEXAS OIL IN THE TWENTY-FIRST CENTURY

"This is not just another cycle we can work out of as we have so many times in the past. The structure of the oil industry is being fundamentally changed, forever."

- John J. Redfern, CEO of Flag-Redfern Oil Company, in 1986.

The petroleum industry in Texas has gone through changes since 1986—changes that have transformed the industry. Some of these changes were responses to price volatility, as crude moved from $36 to $9, then up again and down again. If oil prices are adjusted for inflation, prices moved from $80 per barrel in 1980 (in 2000 dollars) down to $10 in 1998. Every major price decline triggered a new wave of mergers as companies sought economies of scale through growth. Increasingly, the large integrated companies invested their exploration dollars abroad or in the deep waters of the Gulf of Mexico, leaving the surviving independents to pick the bones. In Texas, the industry might well have gone the way of the dinosaur were it not for significant improvements in exploration and production technologies, which offered it a new lease on life.

Part of the "spaghetti bowl" of pipelines between petrochemical plants in the Houston area.

COURTESY OF THE TYRELL ARCHIVES, BEAUMONT PUBLIC LIBRARY.

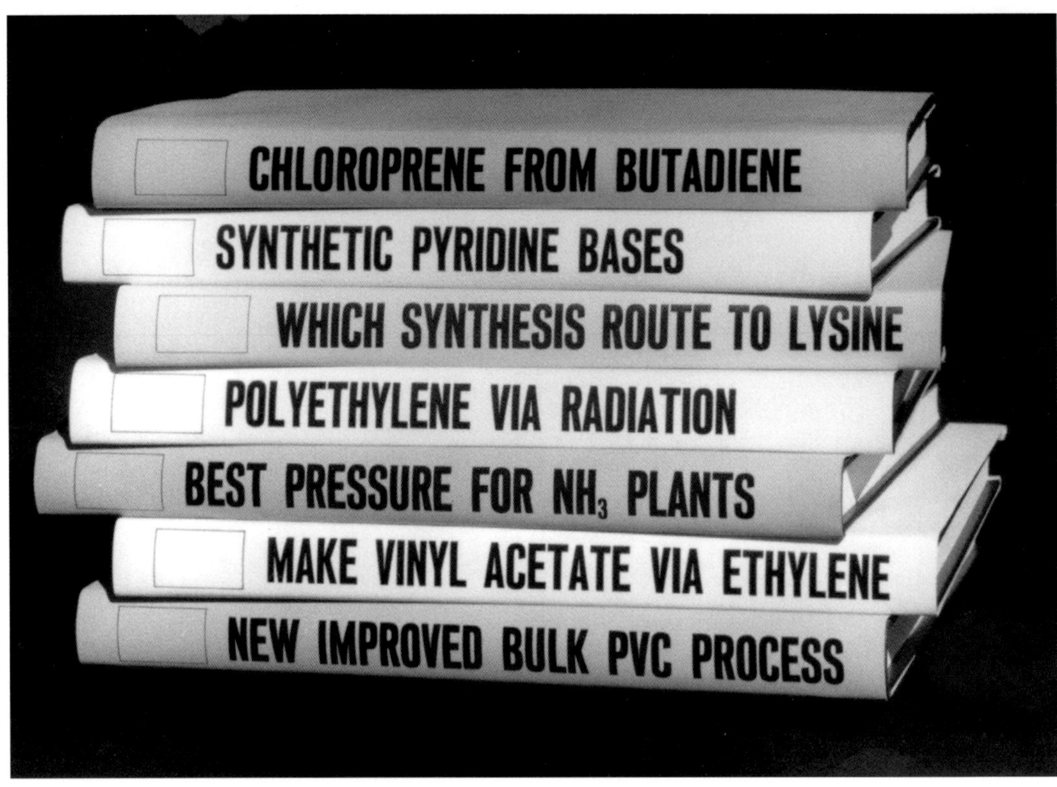

A little light reading for chemical engineers
at a Houston meeting.

Larger independent firms tightened their belts when prices slid sharply. Tom Brown, Inc., for example, sold its sizable Anschutz Ranch holdings in the Rock Mountain Overthrust Belt for $45 million and used the proceeds to reduce its debts. The company also shed its drilling company as rates plummeted, along with its chemicals operation. It wrote down the value of remaining properties by $90 million in 1986 and held on for the long run. Clayton Williams, Jr., another Midland operator, cut his payroll from 1,200 to 200 employees, reduced exploration expenditures from $1.4 million per month to $320,000, and paid off $20 million in debts that carried a 22 percent interest rate. Like the other survivors, Williams later found money to take advantage of opportunities that hard times bought, buying $1-million package of seismic data for $5,000.

Other companies went under. Midland's MGF was stuck with forty-six rigs, most of them idle, went into receivership, and was sold off in bits and pieces. Notably, there were fewer Texas operators, about one-quarter of the number who drilled and produced wells during the boom period. Nationally, the number of independent operators declined from 13,000 in 1984 to about 2,000 in 2001.

The surviving independents generally fell into three groups: small, family-owned companies; scavengers who purchased properties to exploit from major companies and other independents; and strategists who combined exploration and exploitation. The first group, the family-owned companies, worked at keeping costs low—a traditional independent's approach—and rode the tides of changing prices. In Midland, for example, the Wilbanks family continued to operate, drilling two or three wells per year, keeping staff to a minimum, and relying on members of the family to sit on wells and tend leases. The second group, the exploitationists, acquired declining properties from larger operators, invested in engineering and work-overs, and made money by coaxing more oil out of old leases. For example, Titan Resources, a large Midland company, bought Mobil's interests in twenty-seven Permian Basin fields and became a major regional producer, even before it combined its Permian Basin properties with those of Unocal to form Pure Energy Resources. The exploitationists, like the family firms, specialized in domestic operations, and worked to build their natural gas reserves as prices rose beyond the traditional 10-to-1 ratio to crude oil.

The final group, the strategists, including Apache and other company-building independents, expanded through both purchase of assets and exploration. Typically, the growing independents operated both at home and abroad. Hunt Oil, of Dallas, for example, invested in natural gas production, including Canadian properties, and explored in frontier areas, including offshore New Zealand and offshore Newfoundland. Pogo Producing Company, a spin-off of Pennzoil, combined operation of U.S. properties with successful exploration in the Gulf of Thailand. Kerr-McGee acquired what had been the production division of Sun in 1998 and continued to operate and rationalize its relatively large foreign holdings. With sufficient assets, the strategists were in a position to buy properties during price declines and to accept high risks for potentially large gains, even when there were fewer bargains on the table.

As had been true during earlier price dips, some companies combined with others that had deeper pockets and shed properties that were marginal in terms of their rate of return on investment. Thus, Adobe, of Midland, merged with Madison Resources. Parker & Parsley, the large engineering-oriented Midland firm, merged with Mesa Petroleum to form Pioneer Natural Resources. Cross Timbers picked up low-producing leases in the Permian Basin from American Exploration and New York Life Insurance in 1996, drilled additional wells, and increased production. In the first year of the new millennium, Phillips Petroleum and Conoco, two of the leading second-tier integrated companies, combined to create a new company with a market capitalization of nearly $35 billion. In 2003, Matador Petroleum, a Dallas firm, acquired Tom Brown, Inc., and Kinder Morgan, which acquired a large presence in natural gas production, transportation, and processing, bought Marathon's holding in SACROC and in the Yates Field. With every significant change in the prices of oil and gas, there were more mergers and rumors of mergers. The result was a smaller and more durable domestic petroleum industry because companies that were strong enough used the

opportunities produced by low prices and property sales to expand and to strengthen their operations.

Large companies shuffled properties to consolidate assets and shed those that provided the lowest rate of return and reorganized. Mobil acquired Superior Oil, a large producer in Texas. ARCO bought Sinclair then sold ten thousand gas stations and two refineries to British Petroleum, which went on to acquire the Standard Oil Company of Ohio, John D. Rockefeller's original company, and Amoco. ARCO adjusted to low prices by restructuring, creating ARCO Permian to manage and acquire properties and Vastar, another subsidiary that excelled at reworking offshore production, and participated in exploration for new offshore reserves.

As properties changed hands and companies merged or were acquired, mega-independents became major players in the Texas industry. Beginning with Burlington Resources purchase of Louisiana Land and Exploration in 1997, large independents became even bigger. Thereafter, Anadarko, Apache, and Devon added to their reserves through successive acquisitions, with Devon acquiring Mitchell Energy and its large natural gas reserves and processing operations. Anadarko, spun off by Panhandle Eastern, made a series of large buys, including Union Pacific Resources, and built domestic production at the same time that it was participating with Phillips and other larger companies in high-cost projects in the deep waters of the Gulf of Mexico. By 2001, the shuffling of properties had changed the star roster of Texas producers, once dominated by Humble and Gulf, with Occidental Permian emerging as number one, with fourteen percent of the total annual production in Texas.

The list of largest producers also reflected an important change in Texas: some of the biggest actually accounted for small proportions of the whole. For example, the tenth largest, Pure Energy Resources, accounted for only 1.9 percent of total crude production. Not much farther down the roster, Henry Petroleum, of Midland, produced .59 percent of the Texas total. On shore, the majors were gone, except for some natural gas plays, and the typical Texas oil well was a "stripper," yielding less than ten barrels per day. The days of gushers at Spindletop, Mexia, Yates, and East Texas were now in the distant past.

Virtually every merger pushed employment downward in the petroleum industry. Even before it combined with British Petroleum, Amoco had adjusted to declining prices by reducing its workforce by one-quarter. Thereafter, BP Amoco made similar cuts. The combined rosters of Chevron and Gulf fell by a similar proportion between 1984 and 1988. In the service and supply sector, two-thirds of the jobs that had existed in 1982 were gone within ten years. Managers and workers alike had to accept the likelihood that they would not develop careers with a single company. Even those who stayed on did not pronounce the firms' names in the respectful tones of earlier "Humble" men.

Cuts and consolidations continued to cut into the massive refinery-petrochemical operations in Texas. Application of computer controls and other innovations in the highly competitive refineries and petrochemical sectors cut their payrolls by as much as one-third. Texaco and Hoechst decided that the narrow profit margins and the rapid fluctuations in petrochemical markets diminished the value of their installations, which they sold, one plant to Huntsman Financial Corporation, and merged the rest with Celanese. Cities Service got out of petrochemicals completely even earlier, in 1982, when profit margins in the industry fell to -1.3 percent, down from only 3.5 percent a year earlier. In the face of massive costs required to meet clean and water laws, only with relatively cheap natural gas, the principal petrochemical feedstock in the U.S., could Texas manufacturers compete in world markets with foreign concerns that had access to vast supplies in Saudi Arabia and other major producing countries.

Narrow profit margins and the costs of complying with environmental regulations also produced a shuffling of refineries. Chevron sold Gulf's refineries. Ultramar merged with Diamond Shamrock and then was acquired by Valero Energy, a San Antonio company that was already a large-scale independent refiner. After the mergers and acquisitions, Valero owned 13 refineries with a combined capacity of 850,000 barrels per day, had 23,000 employees, and was tied to 5,000 gas stations. It emerged as a larger domestic refiner than Exxon. Other major companies created new corporate ventures with firms owned by foreign nations, as when Kuwait Petroleum and the national companies of Saudi Arabia, Venezuela, and Mexico partnered with U.S. firms in refining, petrochemical, and retail operations. Like the upstream sector, processing and marketing went through a major shift as they lost the competitive advantages that abundant oil and cheap gas had given them.

That cost advantage was lost progressively after the Natural Gas Policy Act changed the marketing of natural gas. The price rose irregularly, rising and falling in line with domestic consumption, which sometimes depended on winter temperatures and summer power needs. By September 2000, however the price had reached $4.26 per thousand cubic feet (mcf), then rose to $6.35 at the end of the year and to $8.06 by January of 2001. By mid-year 2003, the shortage of natural gas was reflected in the decline of gas in storage by nearly one-quarter from five-year averages, and gas prices stabilized at about $6.00 per mcf. The rises squeezed the petrochemical plants—Dow Chemical shifted production from Texas to a plant in Germany. The improved return for natural gas producers had a bright side: it prompted increased exploration for natural gas from one corner of Texas to another. Older areas, such as the Delaware and Val Verde Basins in West Texas, came alive. Major companies and larger independents increased their hunt for natural gas reserves offshore.

Mergers also swept the service and supply sector of the industry. Drilling rates declined steadily with the day rate for a shallow water rig in Texas waters falling from $42,000 per day in 1982 to $8,000 per day ten years later. Adjusted for inflation, at the latter date the rate was ten percent of the boomtime figure. As rig counts fell, equipment was sold, stacked, cannibalized for spare parts, or left to rust.

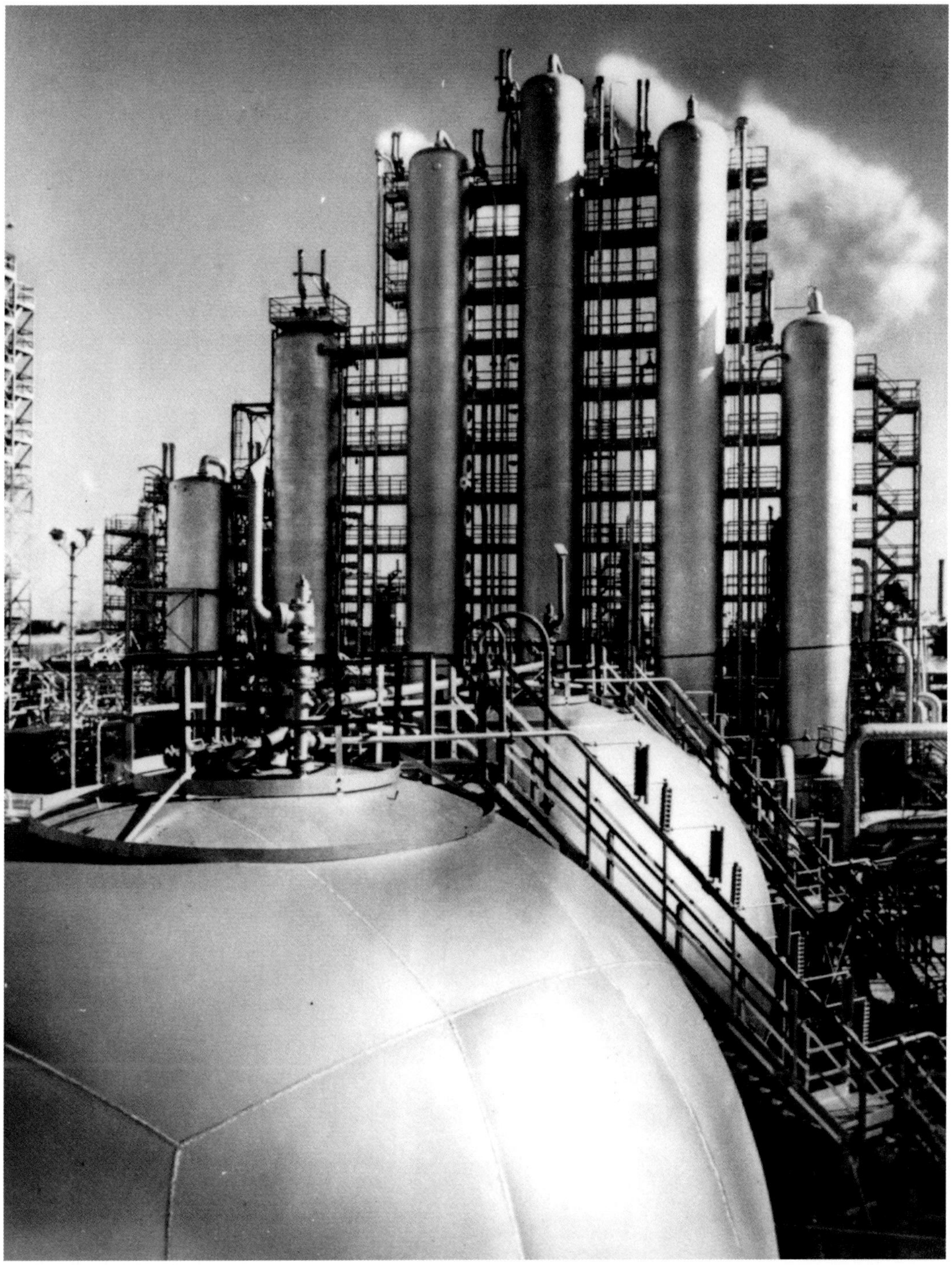

Surviving companies refinanced or sold out. SEDCO, a large drilling company once headed by Texas Governor William P. Clements, Jr., was acquired by Halliburton, which also owned M. W. Kellogg, a principal designer of petrochemical plants, Brown and Root, a major construction firm, and Dresser Industries, a leader in production services. Schlumberger, expanded from its highly profitable drilling services business by buying Houston Industries, which had acquired Western Geophysical, then Schlumberger added Coastal Management Corporation that supplied project management services in North America. Some companies merged just to maintain their competitive positions, as did Baker Hughes in 1986. Others, like Key Energy Services, invested heavily in new technologies and related equipment and focused on the Permian Basin, where its experience and contacts often gave it a competitive advantage. By 2000, medical services and high technology companies both employed many more Texans than the petroleum industry.

A fractionating unit at Exxon's refinery in Baytown.

Within Texas, exploration and production declined steadily. Texas oil reserves dropped by half between 1977 and 2001 with three-quarters of the remaining reserves located in the Permian Basin and West Texas. Production fell at a rate between 10 and 15 percent a year after 1986. Along with the decline, the State of Texas, local school districts, cities, and counties saw their own revenues shrink. At the state level, taxes on various phases of the oil industry had generated a quarter of the revenue. With the bust, two-thirds of that income disappeared. The gusher days were over.

The declining fields, however would still yield oil, especially if engineers installed even more elaborate enhanced recovery systems. The SACROC project, for example, went on from water flood to the injection of carbon dioxide thereby doubling production from the field to the point that it produced more than seven million barrels of crude oil per year. Projects of that kind were costly, and they required careful design and implementation, but the petroleum engineering programs in Texas universities had long prepared students to meet challenges of this kind. Education and capital combined to produce every possible barrel of oil.

Had it not been for the rising value of natural gas and several significant improve-ments in technology, drilling might have declined even more sharply. As it was, significant improvements in seismic tech-nology and in drilling methods made some areas that had been explored for decades without large discoveries more appealing. The development and progressive improvement of 3-D imaging seismic produced shaper images at faster rates making it possible to for geoscientists to identify smaller targets for drilling and to lower the risk of looking for them. The Austin Chalk formation, which spreads from the Mexican border northeastward to the Dallas-Fort Worth area and thence to the Oklahoma border, was worked aggressively with the new data. Some areas, such as the Giddings Field, were found to cover over 1,000 square miles, with pay as thick as 200 to 250 feet of dense and irregularly fractured limestone. Northeast of that formation, the Cotton Valley reef wells became

acceptable risks once the new technology was available. The wells on the reefs were still expensive, costing from $3 million to $3.5 million, but gas production was as high as two million cubic feet per day from them. At that rate pay-out was short and more wells were drilled.

Both the Austin Chalk and Cotton Valley reef plays depended on improvements in specialized drilling techniques as well. Though accidental off-center drilling had long been a problem in the industry and slant-hole drilling had been used legally and illegally onshore and offshore, improvements in measuring and controlling directional drilling made during the 1980s and 1990s had great impacts. When fractures in oil-bearing mineral formations tended to be more vertical than horizontal, a conventional vertical well could only tap a few pays. Horizontal drilling, by contrast, made it possible to drill into multiple pays. The cost of horizontal drilling was often double that of vertical drilling, but in the Cotton Valley reef plays, it often resulted in four times more gas produced per well. In the Austin Chalk, horizontal drilling sometimes boosted production to six times that of wells drilled vertically. With the new technologies, both regions came alive during the 1990s. In the Giddings Field alone, 77 horizontal wells were completed during 1990 and 56 more were permitted by the Texas Railroad Commission. Both technologies also supported in-fill drilling and recompletions in old fields, such as the massive Spraberry Trend. By 2003, about one-quarter of the onshore wells in Texas were drilled horizontally.

The upsweep in natural gas drilling partially offset the decline in oil activity, and it served to boost the importance of the independents who got into it. By 2000, for example, Cross Timbers held two trillion cubic feet of reserves, while Pogo Producing and Mitchell Energy also made the list of top ten independent companies in that sector of the petroleum industry by discovering and/or buying gas properties in the new regions as well as in Southwest Texas and the Permian Basin. Gas, "the new frontier," as trade magazines identified it during the 1990s, offered new challenges and opportunities that Texans were quick to identify and develop. As during earlier periods of growth, new firms emerged to exploit new opportunities.

In Houston, Enron, a natural gas producer, emerged as the nation's leading trader of electricity and natural gas rising to sixty-second place on the Forbes' list of major international firms as it expanded its operations to Latin America and Asia. For a half-dozen years, *Fortune* chose Enron as the most innovative company in America. The company grew to become one of the leading employers and one of the more generous charitable and political benefactors in Houston. As long as the company prospered, it was the ongoing success story in Texas, rivaling Dell Computers and the growing circle of oil majors and independents in town. Its downfall, currently the subject of extensive litigation, has been attributed to high-risk taking and exceptional accounting practices. If nothing else, Enron's rise and fall demonstrated that boom-to-bust has not disappeared from the scene.

In some measure, the controversial Enron story captured headlines while Houston was undergoing a major evolution that seemed less remarkable at the time because it was gradual with relatively few dramatic high points, apart from the merger of Exxon and Mobil. The *Houston Chronicle*, which abandoned its daily oil page when onshore activity declined, had an even bigger story: Houston became what amounted to the world center for exploration. As American companies increasingly sought new reserves abroad, they did more and more business in Houston with partner and service companies. Much as Hollywood had become the most convenient place to make movies, Houston was the easiest place to find other participants in expensive foreign ventures and to secure essential services and supplies. As a result, foreign-owned companies such as Staatoil and China National, and other successful companies that originated abroad, such as Newfield, located on the Bayou, down the road from Exxon, Texaco, Schlumberger, and Halliburton. During the same years, companies increasingly centralized their production offices, moving jobs from Denver, New Orleans, Midland, and other oil centers to Houston. At one time, during the 1990s, the most common complaint in Midland was that "the whole damned world has moved to Houston." By 2001, the move-ins had boosted Houston's oil industry management payroll to boom-time levels and launched Houston into another growth era. Houston's economy was even more tightly tied to oil than it had been fifty years earlier.

The oil age also left Texas with substantial assets including multibillion dollar endowments for its public and private universities. Those institutions typically developed high-quality programs in geology, engineering, and other sciences related to the petroleum industry. The oil industry also fostered the rise of high tech industries. Texas Instruments, for example, was launched by the leading investors in Geophysical Services, Inc.. The legacy of large and pervasive contributions from oilmen and their families has supported world-class cultural institutions in Dallas, Fort Worth, and Houston along with hundreds of charitable institutions across Texas. The industry also expanded the entrepreneurial element of the state's economy with new companies and the optimistic outlook that many oilmen have held to, sometimes in the face of adversity and loss. The major departure from the traditional hostility of farmers toward other businessmen began at Spindletop. Even after massive changes swept the oil industry and the dislocations that followed the cyclical decline of high-tech, Texas is still a place where deal-makers are accepted and respected when they prosper.

The entrepreneurial edge, long-lived and widespread from Beaumont to Borger and Longview to Laredo did not disappear with the most recent boom—and it will not. It seems likely that oil exploration and production will continue to decline significantly every year, and that natural gas production will taper off. But whenever prices offer inducements and oilmen find affordable new technologies or when geophysicists and geologists generate new ideas for them, they will drill again and petroleum engineers will find new ways to coax more black gold and natural gas from the earth.

Suggested Resources

BOOKS

Boatright, Mody and William Owens. *Tales from the Derrick Floor: A People's History of the Oil Industry.* 1970. (An oral history and folklore history of the early industry in Texas, based on hundreds of interviews.)

Brantly, J. E. *History of Oil Well Drilling.* 1971. (The "how it was done" book.)

Clark, James A. and Michel T. Halbouty. *The Last Boom.* 1972. (Still the classic book on the East Texas Field.)

Goodwyn, Lawrence. *Texas Oil, American Dreams: A Study of the Texas Independent Producers and Royalty Owners Association.* 1996.

King, John O. *Joseph Stephen Cullinan: A Study of Leadership in the Texas Petroleum Industry, 1897-1937.* 1970.

Larson, Henrietta M. and Kenneth Wiggins Porter. *A History of the Humble Oil and Refining Company.* 1959.

Linsley, Judith Walker Linsley, Ellen Walker Rienstra, and Jo Ann Stiles. *Giant under the Hill: A History of the Spindletop Oil Discovery at Beaumont, Texas, in 1901.* 2002. (The most complete book on the subject.)

Malavis, Nicholas George. *Bless the Pure and Humble: Texas Lawyers and Oil Regulation, 1919-1936.* 1996.

Myres, Samuel D. *The Permian Basin: Petroleum Empire of the Southwest.* 1973.

Olien, Diana Davids and Roger M. Olien. *Oil in Texas: The Gusher Age, 1895-1945,* 2002.

Olien, Roger M. and Diana Davids Olien. *Easy Money: Promoters and Investors in the Jazz Age.* 1990.

Olien, Roger M. *From Token to Triumph: The Texas Republicans since 1920.* 1982.

Olien, Roger M. and Diana Davids Olien. *Life in the Oil Fields,* 1986.

Olien, Roger M. and Diana Davids Olien. *Oil and Ideology: The Cultural Construction of the American Petroleum Industry.* 2000.

Olien, Roger M. and Diana Davids Olien. *Oil Booms: Social Change in Five Texas Towns.* 1982.

Olien, Roger M. and Diana Davids Olien. *Wildcatters: Texas Independent Oilmen.* 1984.

Owen, Edgar W. *Trek of the Oil Finders.* 1975. (The classic geological exploration book for the region.)

Pratt, Joseph A. *The Growth of a Refining Region.* 1980. (A careful look at the growth of the Texas Gulf coast area.)

Prindle, David F. *Petroleum Politics and the Texas Railroad Commission.* 1981. (The only book-length study of this important agency.)

Spellman, Paul. *Spindletop Boom Days.* 2001.

EXPERIENCE HISTORY

The Permian Basin Petroleum Museum, Library, and Hall of Fame: 1500 Interstate 20 West Midland, Texas 79701. For current exhibitions, operating hours, and entrance fees, consult the museum's website, www.petroleummuseum.org, or call 432-683-4403. Since the museum opened its doors in 1975, hundreds of thousands of visitors have toured its interior and exterior exhibits. Features include working models, re-creations of past geological eras, commissioned oil paintings, and many restored pieces of oil field equipment, including drilling rigs. Guided tours are available. The archives include extensive collections of industry publications, documents, and photographs.

The Texas Energy Museum, 600 Main Street, Beaumont, Texas 77701. For current exhibitions, hours, and entrance fees consult the museum's web page, www.texasenergymuseum.org, or call 409-833-5100. Since it opened in 1990, this museum has become a highly popular site with visitors to Beaumont and for local groups. It includes state-of-the-art exhibits of geology, history, petrochemicals, and refining. A visit may be combined with tours of neighboring art museums and archives, all located in Beaumont's museum district. The fifteen-building recreation of Gladys City, the Spindletop boomtown, is adjacent to the campus of Lamar University.

The East Texas Oil Museum is located on the campus of Kilgore Junior College. The main focus of the museum is the exciting story of the discovery and development of the giant East Texas Field, which it tells through photo exhibits, a professionally produced video, and displays of equipment. The Museum's archives also contain extensive interviews with participants in the boom and a large collection of related photographs. Check the museum's website, www.easttexasoilmuseum.com, for current hours, fees, and exhibits.

The Panhandle-Plains Historical Museum, on the campus of West Texas A&M University, in Canyon, contains the Donald and Sibyl Harrington Wing, which includes a reconstructed oil rig, custom-made exhibits, and displays of historical photographs. The archives of the Panhandle-Plains Historical Association include oil-related items. Check the museum's website, www.panhandleplains.org, for current information.

There are also numerous local museums that include interesting exhibits and materials related to the history of the oil industry in Texas. Local and country historical museums often include exhibits relating to the petroleum industry even when they do not have dedicated oil sites. A partial list of interesting and accessible museums includes the following facilities:

Central Texas Oil Patch Museum in Luling.

The Felty Outdoor Oil Museum and Trails and Tales of Boomtown U.S.A. in Burkburnett.

The Hutchinson County Museum in Borger.

The J. D. Sandefer Oil Annex, Swenson Memorial Museum of Stephen County, in Breckenridge.

The Million-Barrel Tank and Museum in Monahans.

Ocean Star Offshore Energy Center, Pier 19, in Galveston.

The Oil Patch Museum in Batson.

Tower Conoco, a service station restored to its 1930s appearance, in Shamrock.

The Weiss Energy Hall, Houston Museum of Natural Science in Houston.

A restored 1930s Sinclair filling station in Snyder.

The Van Area Oil and Historical Museum in Van.

USEFUL WEBSITES

The New Handbook of Texas, available online, has numerous articles about oil fields and oilmen. It is updated and maintained by the Texas State Historical Association.

The Texas Railroad Commission also posts a large amount of historical information on its website, www.rrc.state.tx.us.

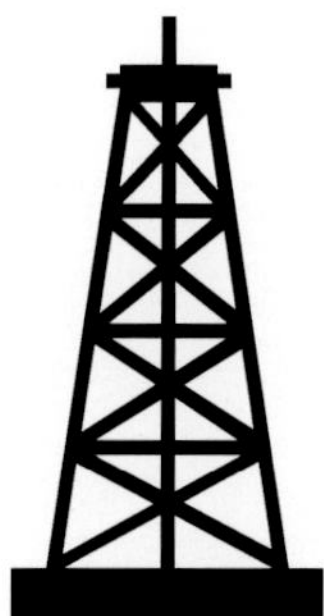

BLACK GOLD

100

SHARING THE HERITAGE

Historic profiles of businesses,
organizations, and families
that have shaped and guided
the Texas oil and gas industry

Dan A. Hughes Company

Dan A. Hughes, Sr., has been involved in the oil and gas industry since birth. His father worked as superintendent of the United Gas Pipeline operation in Palestine, Texas, where Hughes and his twin brother, Dudley, grew up. While in high school, Hughes worked summers on a roustabout maintenance crew on the hundreds of miles of United Gas' natural gas pipelines throughout East Texas. Hughes worked at an adult job even though he was a teenager because World War II had led to labor shortages.

In college, he spent his summer vacations working in the Oklahoma oil fields as a roustabout and other field jobs for Magnolia Oil Company. He graduated from Texas A&M in 1951 with a bachelor's degree in geology. He soon was called into the Army as an artillery officer, where he served in the Korean Conflict on the front lines and was awarded the Bronze Star.

Before going oversees, he was assigned to active duty at Fort Bliss near El Paso, Texas, where he wiled away many off-duty hours mapping the geology of southeastern New Mexico, where an oil and gas play was developing. He purchased a federal lease in 1952 for fifty cents an acre while serving as a lieutenant in the Army and then sold the prospect to a Carlsbad oil operator named George Riggs for an overriding royalty. Riggs later drilled the discovery well for the shallow 650-foot Saladar Oil Field, which is still producing today. After

receiving his Army discharge, Hughes joined Union Producing Company, now known as Devon, and went to work in New Orleans as a geological scout, which was the company-training program for geologists.

A year later he transferred to Beeville and continued his geological scouting. This enabled him to travel all through South Texas, meeting all of the oil operators and drilling contractors, and becoming very familiar with the South Texas oil fields. After being

Above: Dan A. Hughes, Sr.

Below: Woodada Gas Field, West Australia, 1979.

promoted to a full-fledged geologist, Hughes spent several years mapping all of the Cretaceous, Wilcox, and Frio trends of South Texas and doing field evaluation on many wells consisting of coring, testing, and logging. This was an invaluable experience that helped him later in becoming a success in the oil industry and developing hundreds of fields.

In 1961, Hughes resigned from Union Producing Company to become an independent geologist and accepted a retainer from Caddo Oil Company in Shreveport, Louisiana, to consult in South Texas. He realized there were several "old" shallow oil fields in the San Antonio area that had only been partially developed. He recommended the purchase of leases around the possible extensions to these old fields. Operating from Beeville, he picked locations, directed the development, and did the field well evaluations on a massive field extension program, primarily at Bear Creek, Von Army, San Caja, Somerset, and Taylor Ina Fields in Bexar, Medina, Frio, and Atascosa Counties in Texas. Approximately 450 shallow wells were drilled and completed as oil wells based on his geology. Hughes received an overriding royalty for this work, which provided the funds needed to expand the company into larger projects. Following this, Hughes generated a number of prospects in the Frio Trend in the Beeville and Victoria County areas, which were drilled with Flournoy Drilling Company and resulted in several oil and gas discoveries.

In 1965, Hughes formed the Hughes & Hughes Oil and Gas Partnership with his twin brother, Dudley J. Hughes. Dudley lived in Jackson, Mississippi, and worked primarily in the Mississippi and Alabama areas, whereas Dan lived in South Texas and worked the South Texas areas. They combined their resources and began drilling deeper wells, earning a larger share of the working interest because of it. Three private investors provided the partnership with working capital for twenty years. In 1980 the Hughes & Hughes partnership was dissolved. After several name changes, the Hughes' company name became Dan A. Hughes Company.

The first significant South Texas strike for the company came in 1967 with the drilling of the Hughes & Hughes No. 1 Beasley-Connevey in Webb County, Texas. This well, located in a

Above: The Beeville office and employees.

Below: Dan Allen Hughes, Jr., and geophysicist Joe Causey studying 3-D seismic charts.

Dan Allen Hughes, Jr., and Dan Hughes, Sr., at No. 1 Mansker-Discovery Well for the Causey Gas Field, Bee County, Texas, 2000.

remote area north of Laredo, was the discovery well for the ninety-billion-cubic-foot Las Tiendas Gas Field. This shallow field has several gas sands in the interval 2,800 to 3,600 feet and initially resulted in 22 producing Wilcox wells. Later, a deeper development in the seven-thousand-foot level resulted in many Olmos wells over the same acreage. This 11,000-acre field was large enough to entice Houston Natural Gas (now Enron) to build a 114-mile pipeline from Mount Lucas, near George West in Live Oak County, to Las Tiendas Field in Webb County. The Las Tiendas Field served as the basis for continued funding to escalate the company's exploration program. Following this discovery, the company had a continuous series of oil and gas discoveries in South Texas.

The company built its own office building in 1971 in downtown Beeville, but due to the rapid growth of the business, added four additions over the next few years, resulting in a present office building of thirty-three thousand square feet. In considering whether to continue operation in Beeville or leave for a larger town, the company bought its first airplane in 1972. The company found it could operate from this smaller town and fly its personnel anywhere in the Gulf Coast—Texas, Mississippi, or Louisiana—and be back

home the same day, which allowed it to continue its large operation in Beeville. The company has since owned a series of planes that helped contribute to its growth.

In 1970 the company participated in a series of wildcat wells in western Canada with Anderson Exploration Company that resulted in several discoveries of relatively shallow oil and gas fields. The most significant of these was the Dunvegan Field, which turned out to be a 1.6-trillion-cubic-foot reserve. Hughes was fortunate to get into this program through a friend of his, Bob Gowdy, who had transferred in the mid-1960s from San Antonio to Calgary, Alberta, in southwest Canada.

With the success in Canada, Hughes started looking at the possibilities in other foreign countries and took a prospect in Western Australia in 1978. After doing some seismic work and drilling a well, Hughes discovered the Woodada Gas Field located about 175 miles north of Perth. Even though it wasn't a large field by international standards, it was at the right place at the right time, and Hughes was able to sell gas to Perth and the project turned out to be a very lucrative venture.

In 1980, Dan Allen Hughes, Jr., graduated from Texas A&M University and joined the company, spending his first year in Australia developing the Woodada Gas Field. Upon returning to South Texas from Australia, he opened an office for the company in San Antonio. In 1988 the younger Hughes moved to Beeville and joined his father as a partner, and in the management of the company.

Searching for more foreign venture and new areas, in 1996 the company looked to South America for larger prospects. Investigating Bolivia, Peru, and Colombia, the company settled on Colombia as areas with significant potential. Operating as HUPECOL, L.L.C. (Hughes Petroleos De Colombia, L.L.C.), the company has had two oil discoveries, the most recent of which, the Pegita Field, appears to be a significant find.

Dan, Sr., was inducted into the All-American Wildcatters in 1978, and Dan, Jr., became a member of the Wildcatters in 2001. They have both been involved in thousands of wells over their careers. The company has come a long way from the days of Hughes mapping in his spare time while in the Army. They have several 3-D workstations, hundreds of square miles of 3-D seismic data, and generate the majority of the prospects the Hughes' drill. Dan A. Hughes Company is ranked in the top fifty

producers in the state of Texas in both natural gas and oil production.

Both Dan A. Hughes, Sr., and his son, Dan A. Hughes, Jr., plans to continue exploring and wildcatting for hydrocarbons in South Texas and worldwide. A long history of good partners, timing, excellent and loyal employees, and some luck has lead to their continued success.

Above: Hank Kremers, land manager;
Dan Allen Hughes, Dan Hughes, Sr.; and
John Humston, exploration manager,
picking a well location.

Below: A discovery well in Llanos Basin,
Colombia, 2000.

Varco International, Inc.

The first Varco machine shop.

Varco International, Inc., the premier oil field service and equipment company in the industry, is customer-driven.

The corporation has achieved its century of success through a simple philosophy—putting the customers' needs first and developing solutions to impact their productivity and their bottom line.

Following the merger of equals between Tuboscope and Varco in May 2000, under the name of Varco International, Inc., the company now has over 9,500 employees worldwide and generated over $1 billion in revenue during the last fiscal year.

From its leading-edge drilling and inspection technologies to its superior oil field services to its revolutionary use of the Internet, Varco is a fellowship of innovators who specialize in finding smarter, easier, more cost-efficient ways for its customers to maximize their uptime and make their rigs as reliable, productive, and profitable as they can be.

Two Swiss engineers, Baldwin Reinhold and Walter A. Abegg, founded Varco's ancestor, the small California-based hand tools manufacturer Abegg and Reinhold Company, in 1908. Edgar Vuilleumier financed the operation.

Concentrating in the early years on forging hand tools for the western mining industry and the infant automobile industry, Abegg and Reinhold Company furnished the first heat-treated connecting rods, pistons, and crankshafts for racing cars driven by such early greats as Ralph DePalma and Barney Oldfield.

This small company's history in the petroleum industry began in the dynamic times just after the beginning of the twentieth century, with the first methods of rotary drilling. The company's revolutionary method of heat-treating oil field parts enhanced their durability and led to the designation of Baldwin Reinhold as "the father of oil field metallurgy."

Varco, the name they chose for their company (for the names of the three men, Vuilleumier, Abegg, and Reinhold), continues to be identified with many of the most durable tools in the energy industry of the twenty-first century.

The legacy of Tuboscope dates to the 1930s, when Fritz Huntsinger, a German Army veteran of World War I, was operating a small machine shop in California. He was asked by Shell Oil Company to help solve their problem with drill pipe failure in the Ventura Fields. Remembering a device called an optiscope, which had been used during the war to inspect the inside of cannon barrels, Huntsinger ordered two of them. After making modifications, he began using them to detect microscopic defects causing pipe failures. It was quickly apparent that these inspections were saving oil and drilling companies time and money.

New techniques were developed to further improve the service, and separate divisions were created to cover operations worldwide. The Tuboscope process was the forerunner of today's oil field tubular-goods inspection industry.

The merger of Varco and Tuboscope brought together a high-performance equipment manufacturer with a highly responsive and well-known service provider. Inspection and coating of tubulars, solids control equipment and services, and coiled-tubing and wireline products became a part of the Varco product offering. This allowed development of smarter, more cost-efficient ways for the company to minimize its customers' downtime, maximize their uptime, and make their rigs as reliable, productive, and profitable as possible.

In combination, the original industry-leading "firsts" by the ancestors of Varco and Tuboscope form the heritage of an operation that now dominates much of the supply/service segment of today's energy industry. But this was only the beginning of a long line of "firsts" in the history of today's Varco International, Inc.

The corporation has been shaped by the infusion of myriad companies whose talents and technologies have been integrated into a single, synergistic powerhouse. This has been accomplished through a well-planned series of purchases and mergers through the years, each aimed at adding to the list of outstanding products and services.

Each of the individual companies in today's Varco International has a history of its own, many of them dating back to the early days of the twentieth century, when the petroleum industry was at its height of exciting discoveries, and those involved had no concept of the sophisticated production and supply methods that are the modern industry's standard of operation.

What are some of Varco's other "firsts"?

Varco introduced the top drive drilling system in 1982, which revolutionized the drilling process and replaced the traditional rotary drilling table.

By consulting with drillers and operators to improve project economics, it developed the first comprehensive solids-control package to increase efficiencies from flowline to disposal, while ensuring regulatory compliance.

It created the industry's first automated pipe-racking system, which makes and breaks pipe joints, then racks each length quickly and safely, revolutionizing rig operations.

Its production of the first AC-powered Automatic Drawworks System has increased pipe-hoisting speeds while offering greater control and lower maintenance costs than DC-powered units.

In the 1940s, Tuboscope invented the internal tubular coating process, which helps prevent corrosion and increases the hydraulic efficiency of tubing, line pipe, and drill pipe.

Development of Varco's exclusive, "Smart Pig" pipeline inspection tool, another first, lets pipeline owners ensure pipeline integrity. This tool has been used to survey more miles of pipeline than that of any other company.

Varco is number one in the sale and servicing of a host of varying equipment, including rotary-handling tools, pipe racking, iron roughnecks, drilling controls, top-drive systems, motion compensators, pipe transfer equipment, tubular inspections, tubular coating, solids control, coiled-tubing units, coiled-tubing BOPs, wireline units, and mill systems. This product depth, coupled with nearly a century of collective technological firsts and superior customer service, gives Varco a competitive advantage.

The global presence of Varco International includes operations spanning six continents,

Above: An early field tubular inspection.

Below: One of the first mobile tubular inspection units.

Above: The first tubular coating plant in Houston, Texas, in the 1940s.

Below: Tubular inspection.

40 countries, and every major oil field market in the world. Its impressive record of industry firsts translates the company's breakthroughs in product and service technologies into increased productivity, safety, and cost-efficiency for its customers in all phases of the petroleum industry.

Meeting customer needs with innovative products and services is a Varco tradition, but although the company's contributions over the years have changed the way the industry drills for oil and gas, Varco believes that technological firsts have not made Varco the leader in each of its product lines. Its loyal customers have.

These customers cover a wide spectrum of the oil and gas industry, varying widely in size, geographic scope, and technological sophistication. They include operators, drilling contractors, coiled tubing service providers, pipe suppliers and manufacturers, pipeline owners, and refinery operators. They range in size from small service companies and drilling contractors to major multinational oil companies. They do have some common characteristics, however. They want quality equipment that performs as promised and they want service after the sale.

Varco gets their business because the company not only provides these basic needs, but also works continually to exceed its customers' expectations. Varco takes pride in its reputation as the best in the industry in terms of getting the product to the customer. Problem resolution and superior customer service are inseparable oil field needs. Today's customers, who are increasingly focused on financial performance from their shareholders, want products that work the first time, every time, and bring the promised savings quickly to improve their bottom line.

The downsizing trend in the petroleum industry has increased the need for suppliers who can react quickly, with quality service, to fill the gap left by departing employees. Continually changing EPA rules lead customers to seek advice from suppliers' representatives who can offer the experience, available equipment, and service they need, and who can provide performance, reputation, safety, and solutions to their problems. Customers usually call when equipment is down and Varco's 24-hour, 365-days-a-year availability is the first clue to the company's quick, dependable service.

Financial considerations taken into account by customers include pricing that supports a positive cash flow and an appropriate return on their investment, long-term operational reliability, and innovative technologies. Despite constant changes in equipment and operational functions and features in the drilling marketplace, the industry is interested in long-term, not short-term success. This means reliance on good, solid, reliable equipment that performs to their expectations.

In order to accurately determine a customer's needs, Varco representatives make a point of getting out to talk to them. This focus on individual customers provides a much more accurate understanding of the values and needs of each client. Varco's approach to adding value to its services is to establish a relationship, listen to their voices, and then meet their individual needs through technological innovation and world-class service. This has been accomplished through such concepts as assigning corporation representatives to become the customer's voice within Varco and Varco's voice within the customer's organization.

Communications are key to such relationships, with an across-the-board policy of "straight-up honesty" about what Varco can do on the job. Varco believes this policy includes "spending the customer's money like we were spending our own." It all adds up to trust between Varco and its customers.

A philosophy of concern and honesty extends to all the Varco employees who work behind the scenes, from sales to order entry, through engineering and manufacturing, production and assembly, painting, testing, shipping, and invoicing. Since Varco's financial success depends on repeat business, each step is a vital part of doing the job right, satisfying the customers and ensuring customer loyalty.

With its corporate headquarters at 2000 West Sam Houston Parkway South in Houston, Texas, Varco International, Inc., encompasses over 20 divisions in over 40 countries.

A top drive being displayed by Varco at a trade show.

Some of Varco's divisions, and the e-mail address at which each can be reached:

Brandt. . . . . brandt@varco.com
Hydra Rig . . . . hrisales@varco.com
MD Totco . . . . mdtotco@varco.com
Pipeline Inspection
. . . . pipelineservices@varco.com
Pressure Control Engineering
. . . . sales@pce.co.uk
Shaffer . . . . shaffer@varco.com
Elmar. . . . . sales@elmar.co.uk
Texas Oil Tools . . . . totsales@varco.com
Tuboscope . . . . tuboscope@varco.com
Tuboscope Coating. . . . . coating@varco.com
Tuboscope Inspection
. . . . inspection@varco.com
Tulsa Equipment Manufacturing
. . . . tem@varco.com
Varco BJ. . . . . varcobj@varco.com
Varco Systems . . . varcosystems@varco.com.

L. Frank Pitts

L. Frank Pitts, of Pitts Oil Company, has been a leader among leaders in the independent oil and gas producing industry from the beginning of his days in Texas. He remains so today at the age of ninety-four. His energy, vision, perseverance, and inspiration are the foundation of his success over the past sixty years.

Frank Pitts was not born a Texan, nor was he born to wealth. However, his Mississippi farmer and county politician father recognized his son's entrepreneurial ability and nourished this spirit beginning at an early age. His father died when he was a senior in high school. Pitts put himself through Copiah-Lincoln Junior College in Wesson selling "do-it-yourself" automobile paint. Success with the product continued throughout the 1920s and '30s from on-the-road sales, to opening stores across the United States and Europe. His drive and ambition took him from salesman to general sales manager to world sales manager and in 1939 he became president of the Chicago-based paint company.

A few years later, while looking for other investment possibilities, Pitts began investing in the drilling of wells in Illinois and Indiana looking for oil. The wells were dry holes, but that lack of success didn't dampen his entrepreneurial spirit. In 1943, Pitts purchased interests in prospects in Menard, Grimes, and Montague Counties in Texas. The first and second tests were dry, but the third hit oil. He resigned as president of the paint company, formed Pitts Oil Company, and became a full-time wildcatter. These investments would prove to be the "beginning of his passionate immersion in the oil and gas industry that continues to this day."

In 1948, Pitts moved his family from Chicago to Dallas. Pitts leapt into the industry as a wildcatter. "I saw that we had the greatest industry and the greatest economy in the world and I wanted to be part of it." Relying on the most advanced geophysical technology of the day, Pitts started purchasing leases.

Soon after arriving in Texas, Pitts joined the Texas Independent Producers and Royalty Owners Association (TIPRO) in 1949. He then joined the Independent Petroleum Association of America (IPAA) in the early 1950s.

At the time, many independents were vociferous in their criticisms of major oil companies and government regulators. "I thought the TIPRO independents needed to soften their outspoken criticism. Many of the leaders were often hardheaded, hard-driving people, unaccustomed to compromise. Eventually a working group emerged from a broad democratic base that was grounded in an internal dialogue among the activists. This group embodied an enormous range of political opinions as well as personal styles. Gradually this group became a moving force because of the quality each held in common, namely the willingness to hear each other out. And that's the group that I found myself involved with in the '50s."

Above: L. Frank Pitts.

Below: In 1975 to demonstrate the potential areas of undeveloped energy sources in the U.S., Pitts created and distributed to the members of the U.S. Congress and the consuming public, over a million copies of a map of the United States that showed "prospective sediments untouched by drilling" and which was critical to the deregulation of natural gas.

Pitts' vision led TIPRO to become more effective and more focused in its efforts, bringing finesse to the organization, thus making TIPRO more politically effective.

Pitts also developed a network of friendly producers, usually members of state oil and gas organizations from Oklahoma, Louisiana, Kansas, Ohio, Colorado, Mississippi and California. During the 1960s, '70s, and '80s, this group worked together for the improvement or elimination of state and federal laws related to the oil and gas industry. "Over a period of time, we began to accomplish more in our dealings with state regulatory agencies and with the federal government."

In 1960, Pitts purchased the controlling interest of an international geophysical company and became president of Exploration Surveys Inc. (ESI). Through the injection of new marketing ideas, clients included not only the major U.S. oil companies but also others throughout the globe. Nine years after its purchase, Pitts sold the company to U.S. Industries.

In 1970, using the profits from the sale of the geophysical company, Pitts started acquiring thousands of acres of new oil and gas leases. "I had determined that the market for natural gas in Texas was increasing and that it was a time to drill for new production; therefore, my program became rather aggressive for the purchase of leases in Wise and Denton Counties. Shortly thereafter, I pursued similar programs in Parker, Palo Pinto and other counties. There was just one problem: at the time, there were only three natural gas pipelines in the north Texas area— City Service, Lone Star, and Mitchell Pipelines. None of these pipelines were within miles of my drilling activities."

The success of his wells in these Texas counties, but the lack of nearby pipelines, would drive Pitts to create a new standard in the contractual agreement between the Texas producer and gas purchaser.

"In 1973 I was running a number of rigs in Wise and Denton Counties and had already set pipe on many wells. It was necessary to find a market for my gas. I contacted the three north Texas pipelines for a price. The highest price that I was offered was twenty-five cents

per thousand cubic feet; therefore, I decided that I would have a bidding contest among all the area pipelines and any new ones that might be interested in entering the area for intrastate sales of natural gas. (All interstate sales of natural gas were controlled by the federal government since 1954 under a federal court decision in Wisconsin.) The date set was May 4, 1973 and to my amazement there were twelve bids from twelve different companies interested in buying my gas. I had stipulated in the requirements some new things that were not customary in gas purchase contracts. All of these bids included those requirements. In the final analysis the bid that we took was 82.5 cents per thousand cubic feet, including the liquids in that thousand feet."

In the struggle for the deregulation of natural gas, Pitts became a leading spokesman for the industry. He has been interviewed by members of the local and national media for over thirty years, while giving hundreds of speeches to wide and diverse audiences.

In 1973, Pitts changed the Texas independent producer sales contract forever. "Prior to this new gas purchase contract, natural gas in Texas was sold in the thousand cubic feet (referred to as an MCF) contract basis, and whatever liquids contained in the gas stream was given to the purchaser. The producer did not receive any credit for the liquids. My new content contract changed that. Now the producer sells natural gas on a British Thermal Unit (BTU) basis. As an example, if the gas stream had 1250 BTU, the producer formerly would have been paid only for 1000 BTU. Now the producer, royalty owner and state get paid for the full 1250 BTU. In other words, that was a complete reversal of the way natural gas had been sold." As a result of Pitts' business acumen, Texas producers and their royalty owners have since received the full value of their products.

In order to enhance this political effort to deregulate the industry, Pitts and other leaders devised a plan to successfully implement their vision of deregulation. "We proceeded to come to the conclusion that we should have a two-pronged attack of education regarding our needs: one–to educate the consumer, the voter, and, two–to educate the legislators, helping them to become better informed law makers." We sought the support of the voters

through the media and we sought the support of the legislators by conducting office and field presentations.

As an educational tool, in 1975, Pitts and his brother, Shelby, along with 30 other producers and industry suppliers, created a speaker's bureau, which, in that year alone, gave over 1,100 interviews and presentations through radio, television, newspapers, and speeches. The bureau had four people making appointments all over the country to inform the public and seek its influence with national politicians. All speakers volunteered their time and talents during this tremendous drive to deregulate the natural gas industry.

As an example of these efforts, on one day in Atlanta, Pitts completed interviews with the *Atlanta Constitution*, three radio stations, and three national television stations. Pitts appealed persuasively to the American public. "If the consumer wanted more clean and efficient natural gas, they had to encourage the legislators to vote for deregulation. Only this way could the producers increase the price sufficiently to encourage the production of the industry throughout the United States." This effort of education of the electorate, started in the early 1970s, led the producers in the mid-1970s to believe that they had finally achieved their goal of deregulation of the industry with a bill in Congress in 1976.

Right: Pitts has received many distinguished awards over the years for his activities–among them the 1979 Chief Roughneck presented by Lone Star Steel at the annual convention of IPAA.

Below: In 2001, Pitts was given a standing ovation when he received the Texas Railroad Commission Pioneer Award in Austin. Governor Rick Perry sent a proclamation calling him "a true pioneer in the oil and gas industry."

"Although the bill was approved in the Senate, it was lost by two votes in the House. The speakers' bureau members started all over again. Two years later, in 1978, a bill for phased deregulation was passed, and total deregulation occurred in 1985."

Pitts, a tireless advocate of the industry, is known today as "Mr. TIPRO" among the independent producers in light of his "devotion and driving force in TIPRO for two generations."

An activist in his industry, Pitts also made time for involvement in community affairs. He helped found the Dallas Council on World Affairs (now the World Affairs Council of Greater Dallas) and went on to serve as chairman of the board and currently chairman emeritus. He also helped in the formation of the Dallas Opera Company, serving on various committees and is now a trustee. He worked with the Baylor University Medical Center Foundation, serving as chairman of the board, and currently a member of the Executive Committee. Pitts is a current member and past president of Park Cities Rotary Club, a member of Dallas Citizens Council and a lifetime Deacon of Park Cities Baptist Church.

Frank Pitts' industry recognizes and appreciates his outstanding leadership and statesmanship; he has a long list of honors and titles received over the years. In 1979 at IPAA's annual convention, he received the Chief Roughneck Award, the highest honor in the industry. In 2001, Pitts was honored with the Texas Railroad Commission's Pioneer Award for his "sterling contributions to the oil and gas industry." In 2003, Texas Alliance of Energy Producers named Pitts as a Legend Award recipient and Governor Rick Perry recognized him. "You are a Texan second to none. Just as the Texans of days gone by, who worked tirelessly to build a foundation of excellence, you have served with distinction, displaying always the characteristic energy, entrepreneur-ship and vision that have defined a legacy that will live on for the generations to come."

Pitts has participated in over three thousand wells and is actively developing new prospects. "I love this industry and the people in it. I love being in a business where a man's word is his bond. The independents are a great group of people who are finding and producing a product that consumers want because energy is important to this nation."

L. Frank Pitts, the Texas statesman of oil and natural gas, is the "truest independent of them all."

Information from Lawrence Goodwyn's Texas Oil, American Dreams *(Texas State Historical Association, 1996) and from an article by Peggy Williams' in the June 2004 issue of the* Oil and Gas Investor *was used in this profile.*

For many years, L. Frank Pitts was always accompanied by his beloved wife of fifty-eight years, Martha. Martha died in 1993 following a lengthy illness.

H. R. "Bum" Bright

The word "legendary" is often overused when discussing successful Texas oilmen. But if any one person deserves the designation it's H. R. "Bum" Bright, whose work ethic, artful deal-making and tough negotiating skills led him from oil field roughneck to one of the richest men in the world.

Harvey Roberts Bright was born in Muskogee, Oklahoma, on October 6, 1920, to Christopher R. and Rebecca E. (Van Ness) Bright, and it wasn't too long after entering this world that he earned his unique nickname. His father, a salesman for Johnson & Johnson, looked at his red-faced baby wrapped in blankets and remarked upon his resemblance to a railroad bum. The nickname stuck like mud to a drill bit and its use became so common that Bright didn't learn his real name until third grade.

At the age of eight, Bright's parents moved the family to Dallas, where his father eventually purchased University Park Pharmacy. Bright's mother, a former schoolteacher, taught her son and daughter to read and encouraged them to memorize poems, a devotion Bright carried with him into adulthood, when he was known to recite pages of Rudyard Kipling by heart.

Bright graduated from Highland Park High School and went to work as a roughneck in the Texas oil fields, where the days are as long as they are hard. Not to mention hot, dusty, and dangerous, facts that didn't escape Bright's attention. It was on one such day when Bright made an observation that would forever change his life.

He was in the cellar under a drilling rig floor standing waist deep in mud wrestling a blow-out preventer into place when he looked up and spotted a man wearing gabardine slacks and a sports shirt giving orders to others. The man turned out to be the engineer, also known as "The Company Man," and Bright quickly realized he would rather be giving orders than wrestling that big piece of iron.

With help from his parents, Bright enrolled at Texas A&M University in College Station, where he graduated in 1943. On May 27 of that same year, Bright married his high school sweetheart, Mary Frances Smith, and together they had four children—Carol, Margaret, Christopher, and Clay. Mary Frances passed away in April 1971 and Bright married his current wife, Peggy, in December 1972.

Following his graduation, Bright was called to active service in the Army Corps of Engineers, where he served as an engineer in Europe and rose to the rank of captain. After receiving his discharge, Bright went to work for Sun Oil Company, rising to the level of field engineer in less than a year. Although he enjoyed the work, he noticed that his older colleagues often lamented the fact that they hadn't struck out on their own.

Bright decided he didn't want to end up like them, and one day, when he accidentally opened his boss' paycheck, he was convinced that he needed to go it alone. His boss, it turned out, made only $10 more per month than he did and after thinking about it for

H. R. "Bum" Bright.

several weeks, realized that if he ever wanted to make any "real" money he would have to do it himself. He had managed to save $6,500 from his Army pay—a minor miracle when you consider how little they earned—so he decided to quit before children and financial responsibilities made it harder to take risks.

So he teamed up with a former college roommate, Herbert Schiff, and began trading oil leases. Legend has it that on his twenty-seventh birthday he was down to $12.76 in his checking account. But he somehow managed to survive and three years later he and Schiff sold half their holdings for $500,000 and used the proceeds to start Bright & Company, an exploration and production company.

Bright became a millionaire before his thirty-first birthday and as his business flourished. Bright decided to diversify in order to secure his wealth. He ventured into the real estate, trucking, and financial services, and eventually came to own more than 120 companies in a wide range of industries.

Those companies included East Texas Motor Freight Lines, Southern Trust & Mortgage Company, Bright Mortgage Company, Bright Banc Savings Association, Trinity Savings & Loan Association, STM Mortgage Company, Big State Freight Lines, and, in the 1980s, the state's fourth and sixth largest savings and loans, which Bright purchased. He also owned stock in and served on the board of directors for Republic National Bank, Reynolds Penland Company, Southwestern Public Service Company, Republic of Texas Corporation, and Massachusetts Mutual Life Insurance.

Bright eventually owned so many companies that he began to name them after minerals (Dolomite and Antimony), South Sea islands (Sulu and Java), and Italian food (Manicotti and Scallopini). He also chose properties because the address had his lucky numbers—3, 7, 11, 33, and 55.

Bright's idiosyncratic nature and his love for making money drove him to negotiating measures that others would never conceive of much less follow through on. He once followed a man into the bathroom of the Los Angeles Country Club in an effort to make a deal, and talked the wife of a West Texas farmer into locking her husband out of the bedroom until he sold him a five-thousand-acre oil lease.

The fear of being poor also drove him to succeed long after it became apparent that the lack of money would never be a problem. He

kept his family on a relatively modest budget considering his means and long after he had become successful Bright continued to rise at 5 a.m. and be in the office at 7 a.m., where he would work a 12-hour day monitoring Bright & Company, Bright Banc, other business holdings, and the real estate trusts he set up for his four children and 13 grandchildren.

His penchant for hard work and persistence pervaded his office. He coined the phrase "bring back the book" to symbolize his business philosophy. The phrase refers to his days as an A&M student when a professor chastised him for not returning with a library book he had been assigned to get. The professor told Bright that his excuse—that the book had been checked out—wouldn't do, and that the world pays off only on results. Bright, embarrassed by his failure, returned to the library, contacted the person who had checked out the book, and retrieved it.

He learned that persistence pays off, results are the only thing that counts and, if he failed to do his job, he wouldn't get ahead in life. The story became a part of the Bright business legend and employees were known to wear buttons emblazoned with the words "Bring Back the Book."

Throughout his career, Bright kept his debt down rather than take on loans that might come back to haunt him. He paid off his oil company's debt in 1980 and survived the oil bust as one of the few profitable independents. He also kept his companies private, in part, because they don't have to juggle their business to keep stock prices up. Rather than spend time driving up the price of their stock, a private company can concentrate on building a healthy, profitable company.

Bright took few vacations in his life, finding enjoyment in his work instead. He took pride in his achievements and found satisfaction in making his businesses run well and at a profit.

Despite his many successes in business, Bright is known by many as the former owner of the Dallas Cowboys. In 1984, Bright put together an 11-man partnership to purchase the Cowboys from the original owner, Clint Murchison, Jr., with Bright owning the majority interest of 34 percent. The purchase price, $85 million, was the largest in sports history. Bright later sold the team to current owner Jerry Jones.

Making money and deals was Bright's forte, but it wasn't his only passion. He served on the board of directors of Dallas Citizens Council and Children's Medical Center of Dallas. He joined Children's Medical Center's board in 1971 and served as its chairman from 1978 until 1985.

A lifelong philanthropist, Bright gave millions to local charities, including an unrestricted endowment in 1996 of $25 million to Texas A&M University. Bright not only attended the university, he also served the Texas A&M University System as chairman of its board of regents. Bright also contributed time, effort, and money to the Children's Medical Center, including a $5-million donation that allowed construction of the 75,000-square-foot outpatient facility.

His philanthropic efforts, success in business, and reputation as a one-of-a-kind oilman and dealmaker ensures that the legend of H. R. "Bum" Bright will live on in the annals of Texas history.

Bandera Drilling Co., Inc. began in September 1975 under the direction of Ray and Ann Brazzel of Abilene, who ran their fledgling business from the Hartford Street apartment that they shared with their eleven-year-old son, Stephen, and their baby daughter, Kimberly. Ray had worked on and off in the oil business since his sixteenth birthday, and Ann, a former elementary school teacher, set out to learn payroll, accounting, and other office procedures.

Four wells drilled in the fourth quarter of 1975 helped Bandera survive a slowdown in drilling activity in the first quarter of 1976. Work began to pick up by mid-March and, by August 1977, Ray had purchased the company's second and third rigs.

As the company grew, Ray and Ann were running out of space in their living room office. Ray discovered that the old gin property on T&P Lane in Abilene, zoned for heavy industrial, had features that met their needs. They purchased the property in early 1978, had plans drawn up for offices, and moved to the T&P Lane offices in September 1978.

Ray, who had begun attending auctions to pick up supplies, drill pipe, and other items, purchased the company's fourth rig at an Oklahoma City auction when his opening bid, made to help the auctioneer get the bidding started, was the only one submitted.

By March 1980, with the U.S. energy industry in full swing, Bandera Drilling had its fifth rig in service along with nine trucks, operating under Bandera Trucking. As the company grew, Bandera Supply & Manufacturing employed more than 100 welders in the manufacturing of drilling and well-service rigs, while the trucking company expanded to 22 trucks.

Ray and Ann formed Bandera Development in November 1981 with plans to open an industrial development on East Highway 80 named Oil Center. Bandera Drilling remained at the T&P location, while several other companies formed by Ray and Ann moved into new Oil Center facilities.

In May 1982, not long after the company manufactured Rig No. 11, the oil industry entered a tremendous downturn that led to the demise of many oil industry companies. Bandera's growth came to a halt and, as the economy worsened, the company gradually downsized.

Today, Bandera Drilling is a strong company serving West and North Texas with six rotary rigs drilling from depths of 5,000 to 10,000 feet. The company has forged a reputation for safety, service, reliability and quality while drilling more than 3,800 wells in 74 Texas counties.

The company's safety policy, employee training, and strong enforcement procedures are important components in its commitment to the standard: "A job is well done only if it is done safely."

Bandera Drilling Co., Inc.

Above: Bandera Rig No. 11 in Sterling County Texas for Exxon Company USA drilling the Johnson No. 62 well in 1984.

Below: Ray and Ann Brazzel pose for an Abilene Reporter News article covering Bandera's twentieth anniversary.

Hunt Oil Company

It's impossible to tell the story of the Texas oil and gas industry without expending a significant amount of yarn on the role the Hunt Oil Company has played in its development.

The success of the Dallas-based company has, in fact, given it an important place in the growth of the U.S. and global oil and gas industries, and allowed it to establish profitable operations in refining, real estate, power, agribusiness, and private venture capital investments.

It all started with H. L. Hunt, who was born on an Illinois farm on February 17, 1889. He left home at age fifteen to make his way in the world. He worked as a cowboy, lumberjack, and laborer, saving enough money to buy a plantation in Arkansas on the Mississippi River in 1911.

Floods and an agricultural downturn wiped out his investment, but by 1921, Hunt had recouped enough capital to consider buying another plantation when fate intervened. News arrived of the discovery of oil in El Dorado, Arkansas, and stricken with "black-gold fever," Hunt set out for El Dorado, where he began trading oil and gas leases, later drilling on them in Arkansas and Louisiana with growing success.

In 1930, Hunt heard tell that someone was drilling a wildcat well in East Texas, which at that time was not considered a promising area in which to drill for oil. But he traveled to Rusk County, Texas, to see for himself, and met Columbus Marion "Dad" Joiner, a legendary wildcatter who, after two unsuccessful efforts to get a well down, was drilling the Daisy Bradford No. 3. in late July 1930 when the well struck oil.

Hunt recognized the significance of the test results of the Daisy Bradford No. 3 long before the rest of the industry, and, taking a gamble that would become the company's trademark, he purchased the well and nearby leases.

The East Texas Field ultimately recovered six billion barrels of oil, making it the largest oil field in the United States—and the world— at that time. The field provided the financial impetus for the later founding of Hunt Oil Company in 1934 and its subsequent legendary growth.

In the decades that followed, the company continued to discover and develop very

significant oil and gas fields, both onshore and offshore in the Gulf of Mexico in North America. The growing Hunt Oil Company also gave birth to many additional successful companies and also diversified into other lines of business, including refining, marketing and transportation of crude oil, natural gas and petroleum products as well as farming and ranching and real estate developments.

With the passage of time the company matured. It was soon to receive new leadership.

During his high school and college days, Ray L. Hunt worked for Hunt Oil in the field or in the office, learning from the ground up about the oil business. He learned important business lessons from his father as well as from employees at every level of the company. Following his graduation from SMU, with a degree in economics in hand and an appreciation for modern management techniques (tempered by the practical lessons from his work at Hunt Oil Company), he directed his own business operations. But, sooner than he ever expected, he was called on to lead Hunt Oil Company into what would become its second cycle of phenomenal growth.

Following H. L. Hunt's death in 1974, Ray L. Hunt, at the age of thirty-one, became Hunt Oil Company's new chairman. He continued to focus the company on oil and gas exploration and production, but with a significant and very deliberate change in direction. While keeping its domestic operations strong, he quickly directed the company into international exploration. Soon, a remarkable team of geologists, geophysicists, landmen, and other oil finders developed new prospects around the world that represented both risks to the company as well as substantial opportunity. The decision to "go international" proved to be both well calculated and fortuitous.

The company first found international success and worldwide media attention in 1976 when it participated in the discovery of oil in the North Sea, more than eighty miles from the nearest oil production. In an area the industry had long believed would be non-productive, the industry media touted the Beatrice Field as the "discovery of the year" in the North Sea.

Having again discovered oil in an area others disdained, Hunt Oil Company expanded its search for new opportunities abroad. The search led to Yemen, an ancient Middle East nation that most in the industry had written off as devoid of oil despite its location on the oil-rich Arabian Peninsula. Indeed, in 1952 the British ambassador in a letter to the British Foreign Office stated: "As to the possibility of a profitable concession being available near the Yemen border, I think it is most unlikely. It is desperate country once away from the Asir and Yemen highlands, and I should not think it likely to be oil bearing. Moreover, I cannot see any company in their senses seriously considering running a pipeline through the Yemen."

Reunion Tower and the Hyatt Regency Dallas, a Hunt project, is an internationally known landmark for Dallas.

Above: Hunt's operations in South Louisiana are a model for sound practices in an environmentally sensitive area.

Below: Hunt ventured into international exploration activities beginning in the 1970s, first finding success by participating in the discovery well of the Beatrice Field in the North Sea, and later with a major discovery in Yemen. Hunt now has international exploration activities in a number of areas around the world.

In 1981, Hunt Oil Company signed a production sharing agreement with Yemen, and began exploration operations with seismic work and geophysical interpretations. Having identified several prospects, drilling began, and on July 4, 1984, the company's gamble paid off when its first well, the Alif No. 1 rank wildcat tested positive for oil and gas. The well, which flowed at seventy-eight hundred barrels a day, validated the company's optimism and scientific judgments, and would make Yemen, which had previously imported all its oil products, an exporter of oil to world markets.

The discovery of the giant Alif Field triggered a national celebration in Yemen. Less than three-and-a-half years after discovering Yemen's first oil, the technical capabilities of Hunt Oil Company's production team were verified with the completion of a 263-mile pipeline carrying oil to world markets from the vast Rub Al-Khali Desert across the mountains in Yemen to the shores of the Red Sea and then to an offshore floating storage tanker.

Production, which began at 178,000 barrels per day, continues to this day, though at slightly lesser amounts. Hunt Oil Company has discovered approximately one billion barrels of oil and 11 trillion cubic feet of natural gas since 1984. Utilizing modern techniques, the extremely capable production, reservoir, and drilling engineers of Hunt Oil Company will keep production flowing for many years to come.

Through the years, Hunt Oil Company has grown into one of the largest independent oil companies in the world, with core operations onshore in Texas and Louisiana, in the Gulf of Mexico, and Yemen. In the early 2000s, Hunt Oil Company increased its Canadian operations with the acquisition of companies and proper-ties in gas-rich Alberta. In addition, the company greatly expanded its interest in South America with the acquisition of a substantial portion of the Camisea Gas Fields, one of the largest accumulations of natural gas and associated liquids in South America. The fields will deliver gas to Lima, Peru by 2004 and also hold the promise of LNG exports in the years ahead.

Hunt Oil Company is also active in other parts of the world, with onshore exploration or deepwater drilling activity planned in North and South America, Africa, the Middle East, and Asia.

Today, Hunt Oil Company is a substantial part of the group of companies that includes Hunt Refining Company; Hunt Realty Corporation, which invests in commercial real estate developments; Hunt Private Equities Group; Hunt Ventures; Hunt Investment Group LP; and Hunt Power, a utility company that finds opportunities in the deregulating power industry.

Through various Hunt entities, the company's Ranching and Farming Division owns or conducts operations on more than 380,000 acres of land in the U.S. and Mexico, making Hunt one of the most substantial private landowners in the United States. The company's ranching operations include the famous Hoodoo Ranch in Wyoming and the Preston Nutter Ranch in Utah. Its largest farm, the Sharyland Plantation in South Texas, is quickly becoming the biggest commercial and residential development on the U.S.-Mexico border.

A privately held company, Hunt Oil Company has built its reputation by challenging conventional wisdom concerning where to explore for oil and gas, and on making fast and sound decisions as it responds quickly to new opportunities. It takes great pride in its exceptional people who make the company the global success that it is.

Perhaps Hunt Oil Company's reason for success is best found in its mission statement: "To be a growth-oriented industry leader respected throughout the world for the quality and competency of its people, the efficiency and scope of its operations, and its rich heritage of honesty and integrity."

Foy Boyd left King Resources Company in October 1970 to become an independent petroleum consultant and formed Petroleum Consultant Services (PCS). Because it was easier for clients to remember Foy Boyd rather than Petroleum Consultant Services, consequently Foy Boyd Associates was incorporated in 1974.

Foy Boyd Associates, Inc., (FBA) offered petroleum engineering consulting services related to field operations in drilling, completion, production, and property management for independent operators and individuals without technical and field personnel of their own.

Various services included feasibility studies, engineering and design, well site supervision, and property management singly or in an all-inclusive package. Divisions of the services offered included planning, making recommendations, coordinating, expediting, direct supervision, reporting, and evaluating.

FBA was one of the first consultant firms to offer independent oil companies and individuals a complete package of services for the drilling and completion of twenty-thousand-foot-plus-deep gas wells in the Permian Basin.

During the late 1970s and early '80s; FBA was one of the largest if not the largest at that time with 15 full-time employees and 40 associates located in Midland, Texas, and Denver, Colorado, offering consulting services in the Permian Basin and Rocky Mountain areas. Personnel included petroleum engineers, an office support group, and experienced well site supervisors.

Foy Boyd graduated from the University of Houston with a bachelor of science degree in petroleum engineering. His experience and background include drilling and completion experience along with reservoir evaluating with El Paso Natural Gas Company, production experience with Skelly Oil Company and operations manager with King Resources Company.

Foy Boyd Management Corporation was formed in 1978 to generate, screen, evaluate, and manage drilling and production prospects for outside clients and its own account.

FWB Investment was also formed to develop additional investment opportunities related to oil and gas properties, real estate, and other outside potential activities (i.e. Gemini Mud

Logging, American Stabilizer, Atrium Centre Building, and other undeveloped real estate).

Foy Boyd is a licensed engineer in the state of Texas and a fifty-year member of the Society of Petroleum Engineers. He also is a member of the Permian Basin Petroleum Association, Texas Independent Producers and Royalty Owners Association, and the Society of Independent Professional Earth Scientists.

Foy Boyd individually provides expert testimony in oil and gas litigation matters and continues to manage Foy Boyd Associates, Inc., from his office in Midland, Texas.

Above: Deep Hole Consultants in the early 1970s (from left to right): John Hamilton, Leon Gage, J. W. Seales, Kelton Mann, and Foy Boyd.

Below: Foy Boyd with service personnel in well site discussion in the early 1970s.

ENERGEN RESOURCES CORPORATION

BY ROBERT LEE

Above: Among the wells included in Energen Resources' purchase of West Texas oil properties from First Permian, LLC, in April 2002, was this one—the first to be drilled in the prolific Permian Basin.

Below: Energen Resources' aggressive strategy has diversified Energen Corporation in a way that has generated consistent earning growth for the company.

Energen Resources Corporation is a fast-growing oil and gas acquisition and exploitation company with more than one trillion cubic feet equivalent of proved reserves in the continental United States.

The company's operations are focused in four areas: the San Juan Basin of New Mexico, the Permian Basin in west Texas, the Black Warrior Basin in Alabama, and in the North Louisiana/East Texas area. Energen Resources' extensive domestic natural gas reserves place it among the top twenty independent oil and gas producers in America.

In April 2002 the company bought thirty-eight million barrels of proved oil reserves in the Permian Basin of west Texas. It is the company's largest property acquisition to date.

The Permian Basin reserves are long-lived with great exploitation potential, and more than ninety percent of the property's value is operated. The company's reserves in the Texas and New Mexico areas account for more than half of both Energen Resources' total reserves and Energen's net income.

Energen Resources is the largest subsidiary of Energen Corporation, a diversified energy company headquartered in Birmingham, Alabama. In 1995, Energen Resources became the focus of a new corporate strategy designed to enhance long-term shareholder value beyond that offered by Energen's natural gas utility, Alabama Gas Corporation (Alagasco). Under this aggressive growth strategy, Energen Resources became the driver of Energen's long-term earnings growth.

Energen Resources employs more than 250 people and has offices in Houston and Midland, Texas; Farmington, New Mexico; Arcadia, Louisiana; and Oak Grove, Vance, and Birmingham, Alabama.

Energen Resources' history dates back to 1971, when it was formed as a subsidiary of Energen. In the mid-1980s, Energen Resources, then named Taurus Exploration Inc., began carving out an important niche for itself in the field of coal bed methane development.

The company quickly became one of the acknowledged leaders in the development and production of this pipeline-quality natural gas. It managed and operated the investments of nearly $2 billion into the state and created jobs for more than 13,000 people. The company still produces ten percent of Alabama's Black Warrior Basin methane output. In addition, Energen Resources remains one of the top coal bed methane operators in the country.

In the early 1990s the company began to shift its focus outside Alabama toward development of plentiful conventional oil and gas reserves across the United States.

Energen Resources now focuses on increasing its production and proved reserves by acquiring and producing onshore North American oil and natural gas reserves. The company prefers this method to exploration because it minimizes risk

and offers the immediate benefit of existing, proven production.

Energen Resources meticulously analyzes each potential acquisition opportunity. It must be located onshore in North America, have substantial exploitation potential and have a long-lived reserve base.

Over the last eight years, nearly ninety percent of capital spending at Energen Resources has been to acquire and exploit proved reserves. Since Energen introduced its long-term growth plan in 1995, Energen Resources has invested more than $1.1 billion in acquiring proved reserves. In comparison, the company invested $12 million during its first ten years of existence.

At the end of Energen's 1995 fiscal year, Energen Resources had proved reserves of just fewer than one hundred billion cubic feet equivalent (Bcfe) and was producing about 10 Bcfe of natural gas and oil a year. Energen Resources was also contributing about twenty percent of Energen's consolidated earnings. In 2003, Energen Resources has proved reserves of 1.3 trillion cubic feet equivalent (Tcfe), produces more than 85 Bcfe of natural gas and oil and contributes more than sixty-five percent of Energen's consolidated earnings.

In March 2003, Energen Resources decided to expand its San Juan Basin operations by acquiring an additional ninety-three billion cubic feet of gas equivalent in New Mexico and Colorado.

In the areas the company operates, Energen Resources is consistently one of the most active companies in terms of exploitation and operational enhancements. More than thirty years of such successful operation have earned Energen Resources a positive reputation in the industry. The company is respected for its evaluation speed and ability to close a deal. This reputation helps ensure that Energen Resources is invited to review property packages as they become available.

In addition to its reputation as being a productive and efficient operator, Energen Resources is known for its community involvement. The company has a rich history of participation in civic and charitable activities, and many employees are active participants in a variety of volunteer organizations.

Energen Resources is also committed to protecting America's natural environment and resources in all areas where it conducts business. The company's policy is to balance economic development with environmental concerns.

The future looks bright for Energen Resources and its customers. Implementation of the company's long-term growth plan signals considerable investment into the economies of the communities in which it operates. This investment has led to successful diversification and increased production levels, putting Energen Resources in an excellent position to achieve its vision of being one of the best independent oil and gas producers in North America.

Above: Energen Resources is committed to protecting America's natural environment and resources in all areas where it conducts business. The company's policy is to balance economic development with environmental concerns.

Below: Energen Resources is the driver of Birmingham-based Energen Corporation's long-term earnings growth. The company's history dates back to 1971 when it was formed as an Energen subsidiary.

Ratliff Operating Corp.

Ratliff Operating Corp. of Amarillo was founded on March 12, 1984, as Scholl and Ratliff Corp., named for Harvey L. Ratliff, Jr., and Bob Scholl, a geologist who convinced his future partner to test prospects in Nebraska.

Ratliff, who earned a bachelor's degree in petroleum engineering at Texas Technological College, had enough money by November 1972 to allow him to invest. He learned to type and track his business finances and then started investing in drilling prospects, and completing and producing oil and gas wells.

Ratliff and Scholl began investing in prospects in Nebraska around 1984 and eventually decided to form a corporation. But after drilling one producing well and 14 dry holes, Scholl sold his interest in the company to Ratliff, who later changed the name to Ratliff Operating Corp.

After Scholl sold out, Ray Jinkins, the landman who had obtained most of the leases for Scholl and Ratliff Corp., talked Ratliff into drilling his first well in Hardeman County, Texas, north of Quanah that had been dubbed the McGowan 1-A. They completed the well in March 1987 with an initial production rate of 120 barrels of oil per day.

Like everyone in the oil and gas industry, Ratliff learned some hard, expensive lessons through the years. They began in Nebraska and continued in Hardeman County.

The primary objective on the first well was a zone known as the "Palo Pinto Sand." The geologist on the well talked Ratliff into trying a zone about a hundred feet deeper known as the "Hot Shale."

In completing the "Hot Shale," they used acid to supposedly "clean" the perforations and at first thought they had made a wonderful new discovery.

The B-1 well came in at 184 barrels of oil per day, but the D-1 came in at only 25 and the C-1, despite considerable expense, never came in. Ratliff eventually learned that fracturing attempts in the "Hot Shale" had actually gone some 100 feet above into the sand and virtually all the advice he had received from prior operators proved wrong.

Among the misinformation he received was that you must clean the perforations with acid, which actually stopped up the sand and caused all sorts of other problems.

But the expensive lessons Ratliff learned early served him well in the months and years to come. Ratliff Operating Corp. eventually drilled 15 Palo Pinto Sand producers (along with five dry holes) with oil and gas sales of $12,771,612. Between January 1, 1989, and July 2002, investors have reaped $3,430,388.

The company's mission then, as now, was to find drilling locations, drill wells, and complete and produce wells. Through the years, Ratliff Operating Corp. has worked hard to fulfill that mission.

From March 1987 to January 1989, the company drilled three producing wells and one dry hole producing 25,373 barrels of oil

Above: Harvey Ratliff in front of the Steinberger Rig. No. 1, which he used in recent years.

Below: Harvey Ratliff standing by a pumping unit at the McGowan 1A well.

"

from the Palo Pinto Sand in Hardeman County. And from January 1989 to July 2002, the company drilled 23 producing wells and 17 dry holes in Hardeman County, Texas, and Jackson County, Oklahoma.

These wells produced 1,248,821 barrels of oil and 56,377 mcf of natural gas that brought in $24,705,650 in revenues. The company's efforts benefited everyone involved, from investors to subcontractors to the States of Texas and Oklahoma.

From that more than $24 million in sales, the States of Texas and Oklahoma received $1,406,809 in taxes. Royalty owners received $4,701,254, while investors earned $3,864,499. Producing subcontractors or employees received $3,883,477, tangible equipment subcontractors earned $1,931,450, drilling subcontractors earned $8,629,490 and lessors before they were part of the seismic expenses received $288,672.

Harvey Ratliff's sons, Thomas Ratliff and Franklin Ratliff, later played an important role in the company. Thomas earned a master's degree in geophysics from Texas A&M University and worked several years for Mobil Oil Company. In 1992 he decided to go out on his own and his brother, Franklin, a practicing lawyer with a degree in accounting and a CPA

Above: Harvey Ratliff in front of the flanged and flowing Hankins Well No. 1, by far his best well ever.

Below: Harvey and Franklin Ratliff on a derrick floor waiting for a drill-stem test. Billy Green, toolpusher, is in the background.

from the University of Texas and a law degree from Texas Tech University, joined him.

They formed DDD Exploration, Inc., developing 3-D seismic prospects for Ratliff Operating Corp. before branching out to serve other companies as well. As a result of this collaboration—and with the use of 3-D seismic—Ratliff Operating Corp. drilled three deep-zone producers and 11 dry holes. From June 1992 to July 2002, gross oil and gas sales totaled $10,006,099, including $2,555,133 to investors.

Ratliff Operating Corp. has grown from producing 28,922 barrels of oil per year in 1987 to a peak of 120,985 barrels of oil per year in 1998. Cumulative production as of July 31, 2002, had reached 1,317,458 barrels of oil.

The company uses subcontractors for every-thing from pumping to tax preparation. Harvey L. Ratliff, Jr., the company's president, handles the finances, engineering, and business decisions.

It operates in accordance with standard operating agreements between it and its investors, known as "participants" or "non-operators." The company sells its oil and gas to purchasers such as Sunoco, Teppco, and Twin Montana.

Ratliff Operating Corp., a member of several oil and gas associations, has shared its success with the local community through contributions to various charities. Ratliff Operating Corp. plans to continue its involvement in the community while continuing to drill, complete, and produce oil and gas wells.

XTO Energy, Inc.

XTO Energy, Inc., formerly known as Cross Timbers Oil Company, is a premier domestic natural gas producer engaged in the acquisition, exploitation, and development of high-quality, long-lived producing oil and gas properties in Texas, New Mexico, Arkansas, Oklahoma, Kansas, Colorado, Wyoming, Alaska, and Louisiana.

Formed in 1986 and taken public in 1993, XTO Energy is now one of the nation's fastest-growing natural gas and oil producers. The management group, whose members have worked together more than twenty-five years, developed a business strategy that took root during its successful days at Southland Royalty Company and drives the success today at XTO Energy.

The company's time-tested approach to the energy business is built around the practice of buying legacy assets that display a proven pro-ducing history and strong upside potential. The company then enhances the value of these properties by performing low-risk, high-return development activities. This formula continues to grow a valuable franchise. XTO Energy's record of increasing "value per share" consis-tently ranks the company at the top of its sector.

The foundation of the company's strategy resides in its portfolio of properties. XTO Energy acquires the best natural gas and oil properties in the major producing areas of the United States. These properties generally have been productive for decades and are expected to produce oil and gas for many years to come. The advantage of acquiring these long-lived properties is two-fold—a high degree of confidence in the future forecast of production, and the probability that more natural resources exist to develop. Although production declines in all wells, XTO buys properties at a point during the life of the well when the decline is shallow or leveling off. The result is an asset base with one of the longest production-to-reserve indexes in the country. This quality provides a stable, predictable base of production that will deliver healthy economic returns on deployed capital.

The company employs "A-level" geoscience teams to develop hidden upsides using the newest technology, innovative solutions and hard work. XTO has assembled a team of dedicated professionals, including technical staff that average fifteen years of experience, who benefit from the company's efforts to nurture an "intellectual infrastructure," whereby young employees are apprenticed into exceptional performers. This environment has led to low turnover throughout the company's history, while encouraging entrepreneurial thinking. The talented team of employees, now more than 850 strong, has made XTO Energy a top-tier energy enterprise.

The company has an unequaled record of delivering new reserves at the lowest finding costs in the industry. From 1986 to 2002, XTO Energy acquired 2.7 trillion cubic feet equivalent (Tcfe) of natural gas properties and through exploitation efforts has added 2.5 trillion Tcfe to that original amount. This is an increase of more than ninety percent to the original acquired reserves. Also, by operating more than ninety-four percent of the value of its producing properties, the company controls the destiny of development and capital spending. By employing its own people to run and manage daily business, XTO optimizes its extensive operations. And with only five percent of its budget directed to exploration,

XTO Energy avoids high-risk activities. This means the company grows through low-risk events, including workovers, recompletions, development drilling and improving operating efficiencies. As a result, both production and reserves have increased at a compounded annual growth rate of about thirty percent since XTO's initial public offering in 1993.

The combination of all these elements has led to a success story unsurpassed in the industry. The company's operational and financial performances have provided a sound footing for confident investing. Owners are drawn to the company's proven track record, growth potential, low-risk strategy, and shareholder orientation (management owns about 6.5 percent of the stock). For those initial investors who bought stock at $13 per share in 1993, the value has increased to about $125 per share today.

XTO Energy's well-honed strategy has championed a variety of economic cycles in the past seventeen years. With its strong assets and extraordinary talent pool, the company is committed and confident that it will continue to build shareholder values, discover new opportunities and be an investment vehicle leader in the marketplace.

Above: Bob Simpson and Steve Palko purchased the I'll Be Back *bronze by Jack Bryant at a charity fundraiser in the company's early days. It became an important symbol of the company's courage, strength, and competitive spirit. Employees are given smaller versions of the bronze on their fifth anniversary.*

Left: The XTO Board of Directors. Seated (from left to right): Scott G. Sherman and Herbert D. Simons. Standing (from left to right): Louis G. Baldwin, Dr. Lane G. Collins, Steffen E. Palko, Keith A. Hutton, Bob R. Simpson, Vaughn O. Vennerberg II, William H. Adams III, Jack P. Randall, and J. Luther King, Jr.

CLK Company, L.L.C.

CLK Company, L.L.C., a geophysical and geological consulting firm specializing in exploring for crude oil and natural gas reserves in the United States and internationally, was founded in 1974 by R. M. Crout, R. R. Londot and H. G. Kuntz. All three of the founders had worked for 11 or more years for Mobil Oil, where they had the responsibility and authority for all offshore exploration in the Gulf of Mexico. R. G. Gardner, L. F., Burson and C. C. Sorrells soon joined them in their new venture.

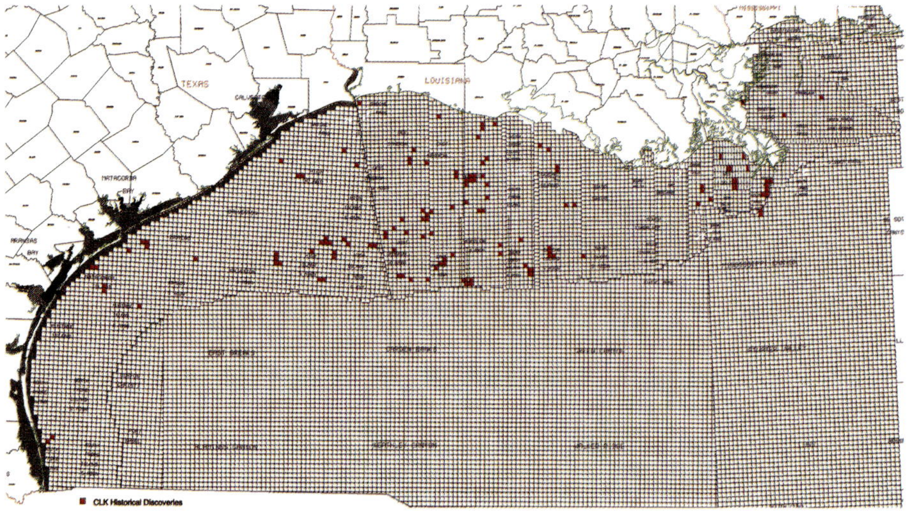

CLK provides exploration guidance through the use of a wide range of geological, geophysical, and computer services, with special emphasis on subsurface mapping, using the latest seismic and well data technology. In certain cases, the direct detection of hydrocarbons is possible through "bright spot" technology. The company's prospect development includes seismic processing, well log analysis, oil and gas reserve volume estimates, risk analysis, and selection of wildcat and development drilling locations.

In CLK's earliest days, from 1974 to 1978, "We were honestly scared to death about survival," Kuntz says. "Our first exploration well was dry. Then we had seven successes in a row. We have been fortunate to have many very qualified exploration people who have joined CLK since that time." He credits a combination of luck and talents for the company's success.

CLK's earliest major activity involved a 1974 exploration analysis of one hundred Federal offshore tracts in the Gulf of Mexico for a group of companies—Ocean Drilling and Exploration, Ocean Oil & Gas, Murphy Oil, General American Oil of Texas, Koch Industries, and Consolidated Natural Gas Producing Company.

Prior to the February 1975 Offshore Texas Federal Lease Sale, CLK conducted an exploration evaluation program involving over six hundred thousand acres in the Gulf of Mexico for another group—Northern Natural Gas (Internorth), Texas Gulf Sulphur, and Total Leonard. This work enabled the group to successfully acquire the lease bids in strong competition with major corporations and other large independents.

CLK found essentially all of Transco's oil and gas reserves from 1975 to 1979, and all of McMoRan's oil and gas reserves from 1977 to the present. Under its contract with Transco, CLK provided the exploration evaluation for the eight Federal offshore lease bids held from 1975 to 1979 in the Gulf of Mexico, completing geological and geophysical analysis on nearly eight hundred tracts. As part of a unified effort, Transco shared CLK's interpretations with The Superior Oil Company, Canadian Superior Oil, Freeport Oil, Louisiana Land and Exploration, Pioneer Production, and McMoRan Exploration. The success of these efforts was reflected in the significant position held during that period by Transco and its partners in producing oil and gas from the Gulf of Mexico.

Beginning in 1977, CLK became the exclusive exploration consultant to McMoRan Exploration in its Gulf of Mexico farm-in program. After McMoRan and Freeport merged in late 1981, the association with CLK expanded to include the onshore United States. CLK's success is apparent from its record of successful exploratory wells. In 1988, for example, Freeport-McMoRan Energy Partners had a 79 percent success rate on the 34 exploratory wells they drilled onshore and offshore in the United States. They achieved a 97 percent success rate on the 131 development wells they drilled that same year.

In February 1992 some of Freeport-McMoRan's senior exploration management joined CLK as partners and associates, and CLK's exploration responsibilities expanded to include all international and domestic exploration. In addition to geological and geophysical services, CLK will continue to provide all of the geological and geophysical exploration evaluations for McMoRan Oil & Gas Company on a worldwide basis through a consulting retainer agreement.

In order to participate in a working interest capacity with McMoRan and other operators, CLK has also formed three separate companies—CLK Exploration Company, CLK Producing Company, and CLK Oil and Gas Company.

Two of the founders are still active in the company. Kuntz has served as president and managing partner since CLK began and is responsible for the company's overall effort of evaluating potentially economic oil and gas prospects. A San Antonio native, he received his BSE in geological engineering from Princeton and an MBA from Oklahoma City University. Londot, a CLK manager, was born in Chicago, Illinois, received his master of science in geology from Louisiana State University and continued graduate work in geology at the University of Texas.

Gardner and Burson also remain active in CLK. Gardner, a professional engineering graduate in physics from the Colorado School of Mines, joined CLK as manager of geophysics, and is executive vice president. Burson, who received a professional engineering degree in geophysics from the Colorado School of Mines and studied petroleum engineering at the University of Texas, joined CLK as a director and vice president.

With offices at 1001 Fannin, Suite 777, Houston, Texas, and at 1615 Poydras, Fifth Floor, New Orleans, Louisiana, CLK has a staff of six technical personnel—geologists, geophysicists, and computer experts—with a support staff of four.

Great relationships with individuals and companies, such as Transco Chairman Jack Bowen, Freeport Chairman Benno Smidt, and McMoRan Chairman J. R. Moffett, and CLK's first close relationship, with "Doc" LaBord, chairman of ODECO, originator of the first movable offshore drilling rig, have been vital to CLK's success. CLK plans to continue utilizing the skills that have created success over the past twenty-eight years, while embracing new technologies and innovations in successful exploration.

Actively involved in community and charitable causes, CLK participates in the Houston Partnership and Houston Chamber of Commerce. Kuntz serves as a trustee for the Houston Grand Opera, Symphony, and Ballet, and CLK and its personnel contribute to almost all significant charities and events.

Cumulative Production by Field Through September 2001

Area/Block/BCFGE

Fields Greater Than 200 BCFGE

VR 25	291.91	VR 225/226*	201.84
MU A-85*	219.33		

Fields Between 150 and 200 BCFGE

HI A-446/A-447/A-448/A-438*	186.92	EI 308/315*	156.93
WD 88/89*	180.97		

Fields Between 100 and 150 BCFGE

VR 310*	145.69	SP 74/82/83*	126.70
WC 498*	143.07	MI 527/528/555*	125.14
EC 331/332*	142.51	MI 700/713*	106.99
WD 104/105*	139.25	WC 215*	104.79
SMI 173/174/175/176*	137.73	VR 143/159/160*	103.30
VR 58	136.65		

Fields Between 50 and 100 BCFGE

VR 146*	96.41	SP 45/46*	62.46
HI A-551/A-552*	85.57	WD 34*	61.36
MC 365*	85.15	HI A-471*	61.29
VR 161/162*	67.90	VR 409/410/389/EC 362*	56.26
EC 143/148*	66.20	EC 311	53.94
SA 3*	65.05		

Fields Between 20 and 50 BCFGE

VR 167*	44.85	WD 62/63*	31.02
EC 336	39.52	GA A-126/A-131	29.11
EC 263	38.23	WC 231/240*	28.33
SM 161*	38.20	SP 54	25.88
EI 97/108*	38.17	WC 464*	24.71
WC 433	34.26	HI A-442*	22.57
HI A-492	33.63	EI 329/330*	22.47
EC 320*	32.35	HI A-284	20.74
EI 191	32.02		

Fields Less Than 20 BCFGE

WC 548	19.86	SM 114*	6.42
VR 36	19.27	SS 296*	6.12
HI A-289/A-290	18.77	VR 107	6.07
WC 457/458/459	18.74	EI 193*	5.70
EI 46	18.30	VR 117	5.65
WC 510*	18.26	WC 236	5.34
WC 616-617*	16.98	VR 196	5.25
GI 79	16.34	GI 83*	4.79
GI 58/65*	16.15	VR 202	4.75
WC 503*	14.19	HI 164	4.24
WC 628	14.16	HI A-178	4.10
HI A-476	13.84	EC 226	3.43
EC 214	9.72	BA A-19★	2.11
HI 262	8.63	VR 28	1.60
VR 332	8.05	WC 492	0.74
EC 26	7.96	GA 211	0.06
GA 156/157	7.71	WD 26	0.04
EC 200	6.56		

GRAND TOTAL: 4267.00 BCGFE

* Still producing.

★ Recent discovery; first well producing 120 MMFCG/D.

Ellison Miles on a work site in the late 1960s.

Ellison Miles is responsible for much of the success the Texas oil and gas industry enjoyed in the last century. This native of Liberty, Texas, had a hand in two of the most significant strikes in the state's rich exploration history.

But Miles, who attended Texas A&M University and graduated from the University of Houston with a degree in petroleum geology, enjoyed a taste of success long before striking it rich in the oil and gas fields.

His first great achievement occurred as a highly decorated B-17 pilot in World War II. He earned the Distinguished Flying Cross for his efforts flying missions over Europe during the war. He even survived the crash landing of a plane that had been wounded over Germany, limping home with an engine riddled with gunfire.

His career as a successful wildcatter began in East Texas with his father-in-law, A. O. Phillips of Dallas. He formed Miles Production Company in 1946 and in 1949 organized Trio Drilling Company.

Miles and Phillips operated a successful exploration company that became even more successful in the late 1940s. That's when the two men—along with A. D. (Bud) Lee, an importer and construction company owner—discovered one of the richest oil fields in the state in Smith County at the A. O. Phillips Staples No. 1 between the Hawkins and Van Fields.

The strike caused investors to flood the area and offer the then-unheard-of-sum of $1,000 an acre for oil leases. The find further solidified Miles' oil and gas prospects and allowed him to move operations to the Dallas area.

But it was an Englishman who would unwittingly lead Miles to his next great success. Colonel D. J. (Jimmie) Hughes owned a twenty-eight-hundred-acre ranch near Boonsville in Wise County, Texas. He had somehow developed the notion that a river of oil flowed beneath his ranch and he wanted someone to find it for him.

So, in 1948, Hughes contacted friend and geologist John A. Jackson. Hughes kept after Jackson to survey his land and "tell me where the oil is." Jackson tried but when he drilled a fifty-eight-hundred-foot-deep hole he came up with natural gas instead. Other wells drilled in the area produced the same results.

Miles and Jackson eventually decided to combine forces and drill the D. J. Hughes No. 1 in 1952 on a lease given to Jackson for past services to Hughes in what was once described as "Frustration Field." The natural gas field discovered there turned out to be the third largest in U.S. history. Miles served as the drilling contractor for the project. He brought in the first well with an estimated reserve of more than five billion cubic feet.

The only problem was that there was no place to sell the gas because there was no pipeline to transport it. The field had to be abandoned for more than a year while a $32-million pipeline was built to transport the natural gas from Wise County to the main line of Natural Gas Pipeline Company of America. The Federal Power Commission approved the sale of 105 million cubic feet of gas daily in the Wise County area to the pipeline company, which sold it to the City of Chicago.

Christie, Mitchell, and Mitchell of Houston (which later operated as George Mitchell & Associates) bought the well and Miles and Jackson served as drilling contractor and geologist. The next two years they leased 285,000 acres in the area and brought in 110 gas wells and 57 oil wells. They averaged ten productive wells for every dry hole.

Miles reorganized Miles Production Company in 1959 with him as sole owner. He eventually drilled more than one thousand wells in Wise County and became known as the "Wizard of Wise County." In 1964, Miles bought the interests of his associates in Trio Drilling Company and changed the name to Emco Drilling Company with Miles as sole owner. Then, in March 1970, after completing the drilling of its 1,055th well, the company was dissolved and the equipment sold at auction.

Although the drilling company no longer existed, Miles Production Company continued on. Miles sold his drilling operations in Wise County to George Mitchell & Associates for overrides on the more than one thousand wells drilled in the county.

He also made a gentleman's agreement with Mitchell not to work in Wise County and after the sale moved to nearby Jack County to explore. His geologist was Ann Stanley and the company continues to operate the wells discovered there.

Since 2000 the company has participated with other companies in drilling operations in

Young, Montague and Smith Counties. The Smith County work brought Miles full circle, to the place where he and his former father-in-law first struck it rich. Miles, at eighty-five, continues to come into the office on a daily basis.

His story is recorded in the annals of the oil and gas industry. His legacy will live on in the Ellison Miles Geotechnology Institute at Brookhaven College. The Texas wildcatter donated $3.5 million to create the institute, which is one of the premier full-service training and conference facilities in the Dallas/ Fort Worth area.

Above: Elison Miles in the late 1970s.

Below: Ellison Miles with memorabilia from his days as a B-17 pilot during World War II, circa 1999.

Glassell Producing Company, Inc.

Alfred C. Glassell, Jr. on drilling rig No. 3 in south Louisiana in 1937.

Alfred C. Glassell, Jr., consultant of Glassell Producing Company, Inc., has taken a lifelong interest in the petroleum industry, in which he is today one of the Gulf Coast's leaders. Born on a cotton plantation near Shreveport, Louisiana, on March 31, 1913, he is a son of Alfred C. Glassell, Sr., and Frances Lane Glassell. His father was an illustrious oil industry figure whose activities date back to the industry's early days.

The younger Glassell received his bachelor of arts degree from Louisiana State University in 1934. His consuming interest in people was evident even in his college years, when he served as student body president, a member of thirteen honor societies, a president of Kappa Alpha fraternity, and commanded the Honor ROC Battalion.

During summer vacations and immediately upon graduation, he began working for his father as a roustabout in order to learn as much as possible about oil field drilling operations from the derrick floor up. Utilizing his experience with drilling rigs and several years spent traveling throughout Louisiana, Texas, and other Gulf Coast areas as a scout for oil companies, he formed the Glassell Drilling Company.

This company was very active in drilling operations for deep wells on the Gulf Coast of Louisiana and Texas. Glassell was personally instrumental in the extension of the Jennings Field in Acadia Parish and discovered the Anse La Butte Field in St. Martin Parish. He also extended the Port Barre Field in St. Landry Parish; the West Grand Lake Field, Cameron Parish; and the Bayou Boullion Field, Iberville Parish, all of which are Louisiana locations. In Texas, Glassell's company discovered east flank production of the Moss Bluff Field in Liberty County and a giant gas field in Carthage, Panola County.

During World War II, Glassell entered the armed services as a second lieutenant, saw active service in the African and European theaters and was honorably discharged with the permanent rank of major.

Upon return to civilian life, his marked ability as an organizer led to leadership in numerous organizations and committees, for which he has served as a director and committee member. He served for many years as director of the Independent Petroleum Association of America. Espousing a credo of free enterprise and individual effort, he has been a strong champion of the independents ever since entering the oil and natural gas business. He was one of the small groups of men who drew up plans for a large gas transmission system to the East Coast. As a result of these plans and as a member of this group, Glassell was one of the founders of the Transcontinental Gas Pipe Line Corporation, which presently operates a transmission system from Brownsville to 132nd Street in New York City. This was one of the biggest pipeline construction jobs of its kind anywhere in the world.

In addition to his work as an independent oil and natural gas operator, Glassell is a former director of the Independent Petroleum Association of America; a director of the Texas Independent Producers and Royalty Owners Association, Down Town Real Estate Company, Spokane Gas Fuel Company, Pacific Northwest Pipeline Corporation, First City National Bank of Houston, Transcontinental Gas Pipe Line Corporation, and the Texas Radio Corporation; a trustee of the Museum of Natural History of Houston; a former director of El Paso Natural Gas Company in Houston; and is currently chairman emeritus of the Houston Museum of Fine Arts.

Glassell is also internationally famous as one of the foremost big game anglers in the world and pioneered the waters off the west coast of South America. The Texas angler has fished the waters off East and West Coasts of the United States, the Bahamas, Mexico, Indian Ocean, Bay of Panama, and New Zealand, and has captained the United States Tuna Team. He set the world record for the largest marlin ever caught on a hand-held rod and reel. At 1,560 pounds, the record remains today.

He is a member of the Fishing Hall of Fame and the International Game Fish Association of Florida; and holds membership or former membership in the Atlantic Tuna Club, of Providence, Rhode Island; the Boston Club of New Orleans, Louisiana; the Cabo Blanco (Peru) Fishing Club; the Shreveport Club; the Texas Corinthian Yacht Club of Kemah; the Bay of Islands Swordfish and Mako Shark Club of New Zealand; Sportsman's Club of America of Chicago, Illinois; The Explorer's Club of New York City; and the Bachelor Club, The Cork Club, River Oaks Country Club, The Coronado Club, the Petroleum Club, the Ramada Club, the Houston Gun Club, the Houston Country Club, and the Allegro Society, all in Houston.

Glassell is associated with numerous societies and museums, including American Geographical Society and American Museum of Natural History of New York City; the Smithsonian of Washington, D.C.; the Museum of Natural History, the Contemporary Arts Museum, the Museum of Fine Arts, the Art League, the Allied Arts Museum, and the Glassell School of Art, all in Houston.

His business address is that of Glassell Producing Company, Inc.: One City Centre Building, 1021 Main Street, Suite 2300, Houston, Texas 77002.

Above: The discovery well of the Carthage Field in East Texas in 1936.

Below: Alfred C. Glassell, Jr., and son, Alfred C. Glassell III, in Glassell Producing Company's Houston office in 2002.

James Cleo Thompson, Jr.

James Cleo Thompson, Jr. was about to enter the law school at Southern Methodist University in 1952, when his father bought three small, rundown oil leases near Crane, Texas. Thompson began cleaning up the leases and putting five wells there back into production. This experience stirred a fascination with the people, business and values of the oil and gas industry, leading to an outstanding record of achievement that continues today.

Combining enthusiasm, business ability, and a code of ethics that makes certain every deal is fair to all those involved, Thompson embarked on a professional career that has included independent oil and gas production, ranching, banking and investments. He is the managing partner of J. Cleo Thompson and James Thompson, Jr., L.P., with properties located principally in Texas, New Mexico, Colorado, and Oklahoma; the managing partner of Double "T" Ranch, L.P., which owns and operates ranches in Denton, Crockett, and Hood Counties in Texas; and the president and chairman of the board for Thompson Petroleum Corporation, also engaged in the oil and gas exploration and production business.

In addition, he has served in numerous business-related capacities. He has been a board member for the National Center for Policy Analysis since 1999; director of Texas Capital Bancshares since 1999; director and a member of the Loan and Discount Committee of NorthPark National Bank in Dallas from 1964 to 1989. From 1983 to the present he has been the board chairman and founding stockholder of Crockett National Bank in San Angelo and Ozona, Texas.

In 2002, Thompson was appointed to the National Petroleum Council, advisory committee for the U.S. Secretary of Energy. Other participation in organizations of the oil and gas industry has included membership in the Independent Petroleum Association of America, for which he has been a member of both the Natural Gas Committee and the Roustabout Club; as well as membership in both the Texas Independent Producers & Royalty Owners (TIPRO), and Hard Hatters.

The recipient of numerous awards for professional and community service, Thompson received the Lone Star Steel "Chief Roughneck Award" in 2001, considered the most prestigious honor in the petroleum industry. This award acknowledges Thompson's lifetime achievement in that field, which includes participation in more than eight thousand wells. He was also the recipient of the Hard Hatters Award for "Outstanding Work in the Oil & Gas Industry" in 1998.

Thompson has also been honored by educational institutions for his outstanding achievements and contributions, with awards including an honorary doctor of letters degree from Howard Payne University in 1991; the

Distinguished Alumni Award from SMU, the highest award the SMU Alumni Association can bestow upon alumni of the institution, in 2000; Honorary Distinguished Alumni from the SMU Law School in 2002; and most recently, the Distinguished Alumni Award from Highland Park High School of Dallas.

Crediting Southern Methodist University with having given him the tools he needed for success, explaining that he learned not only in the classroom, but also on the football field, as a member of the famed SMU teams of the early 1950s.

Even so, he claims that his greatest accomplishment at SMU was meeting and marrying Dorothy Kuntz (SMU 1953). Their daughters, Linda Thompson Gordon (1976) and Christy Thompson (1991), share their parents' alma mater, along with Thompson's love of the petroleum industry. They are actively participating in the family business.

Thompson, who retains a fierce loyalty to SMU, served on that university's board of trustees and its investment committee from 1984 to 1988; and established the James Cleo Thompson, Jr., Professor of Law Chair at SMU in memory of his father in 1984. He also established an endowed Thompson Law Scholarship program in 1984. This program has awarded 80 full-tuition three-year law school scholarships with five scholars per year level.

Also at SMU, he has been an advisor for the Maguire Energy Institute, a post that involves review and evaluation of prospective programs, with board members responsible for contributing ideas on critical issues facing the industry, providing contacts with other industry leaders, and supporting the Institute's marketing activities.

He is an active member of the President's Roundtable "Partners for Tomorrow" Campaign, the SMU Lettermen's Association, the SMU Alumni Association, and the Mustang Club, and has funded the coach's conference room at Gerald Ford Stadium.

As chairman of the board for the Hatton W. Sumners Foundation in Dallas since 1975, he has participated in the foundation's acti-vities, including administration over 150 full law/government scholarships at 13 colleges and universities. Under Thompson's chairmanship, the Foundation's assets have grown over 3,000 percent and the Foundation has awarded more than $39 million in grants and scholarships.

Thompson is also a member of the board of trustees for Howard Payne University and of the board for HPU's Douglas MacArthur Academy of Freedom; a past member of the board of stewards for Highland Park United Methodist Church; a member of the board of directors for the Permian Basin Petroleum Museum; and a member of the Dallas Council on World Affairs.

James Cleo Thompson at a 2001 Houston IPAA meeting moments after being named "Chief Roughneck" for 2001.

ATOFINA Petrochemicals, Inc.

ATOFINA's Port Arthur Refinery traces its roots to the 1930s when Atlantic Refining Company, which already operated a major crude oil terminal at the site, saw an opportunity to increase profits by converting a portion of oil to gasoline.

Located on the Neches River in Port Arthur in Jefferson County, the refinery is officially known as ATOFINA Petrochemicals, Inc. Based in Houston, ATOFINA Petrochemicals, Inc. produces petrochemicals, plastics, intermediates, and polymers in Texas and Louisiana, and in Port Arthur owns and operates an integrated refining complex. This complex produces gasoline, diesel, jet fuel, propane, butane, bunker oil, and asphalt, and recovers benzene, toluene, and xylene—the basic building blocks for plastics—from its fuel products.

Atlantic Refining Company opened the Port Arthur Refinery in 1936 with a visbreaker, a predecessor to modern gasoline production units. In 1944 the company expanded its ability to produce motor fuels by adding a fluid catalytic cracking unit and added an alkylation unit in 1956, and a reformer unit in 1957. The company opened a natural gas processing unit in 1959, and added another crude unit in 1962.

Atlantic merged with Richfield Oil Corporation in 1968 to form Atlantic Richfield Company (ARCO), and less than a year later, BP purchased the Port Arthur refinery, which it owned and operated for four years before ATOFINA bought it in 1973.

ATOFINA transformed the refinery from a relatively simple operation into a sophisticated facility known for the latest technology and top-rate efficiency. In 1973 the refinery processed fifty-four thousand barrels of oil per day, ranking it in the bottom quartile among its peers for efficiency and cost per barrel.

Today, the Port Arthur refinery, considered one of the country's most modern and efficient refineries, is in the top quartile in many categories and its capacity has grown to 240,000 barrels per day, making it the twentieth largest refinery in the United States.

In 1973, ATOFINA reactivated and revamped a crude unit and increased throughput capacity to eighty-five thousand barrels per day. It began a two-phase expansion in 1982 that included construction of a resid solvent extraction unit, a continuous catalytic reformer, an isomerization unit, a benzene/toluene/xylene unit, a sulfur recovery/SCOT unit, and a hydrodesulfurization unit. This modernization replaced older processing units with more modern equipment, which improved refinery efficiency, and product yields.

ATOFINA began another modernization and expansion in 1988 to lower operating costs and increase product yield and throughput. This included a new atmospheric crude unit, amine treating unit, saturate gas liquids recovery unit, and fluid catalytic cracking unit.

The Port Arthur refinery is now designed to run a high percentage of low-cost, heavy, sour crude. Most of its processing units have been built since 1984, incorporating excellent technology and providing a balance between crude capacity and major downstream units.

Yet another major expansion concluded in 2001. In partnership with BASF, ATOFINA built the world's largest naphtha cracker at Port Arthur to produce feedstocks for ATOFINA's plastics operations near Baton Rouge and Houston using naphtha manufactured at the refinery.

Construction on the $1-billion facility took about three years and started up in 2001 as the refinery evolved into a world-scale petrochemical complex that sits on 1,244 acres of developed and undeveloped land. ATOFINA has now joined BASF and Shell in another partnership to build an integrated C4 olefins complex at the refinery. Construction of this $250-million facility began in 2002 with startup in 2003.

The Port Arthur complex places great emphasis on safety, employees, and community. The site safety record is in the top quartile of refiners in the Gulf Coast, which are known as the best of the best.

On-site personnel receive sixteen hours of safety training annually. ATOFINA has replaced traditional CPR training with the American Heart Association's Heartsaver AED course designed to coordinate CPR and automated external defibrillators.

The company, which employs approximately 435 people, also has adopted a Universal First Aid Program that provides in-depth, hands-on skills for participants who may encounter emergencies with delayed emergency medical response.

The ATOFINA Port Arthur Refinery is greatly involved in the local community and each year adds new projects to its program of support. The refinery's Community Involvement Committee uses well-established criteria to make its decisions on whom to support and an ATOFINA employee champions almost all projects.

The refinery's Minority Community Action Committee supports minority programs in Port Arthur. The committee each year provides four full scholarships to minority graduates of Port Arthur to attend LIT's process operator school. The committee also mentors the scholarship recipients to help them through the two-year program.

The ATOFINA Port Arthur Refinery participates in a wide range of industry trade groups and is committed to environmental stewardship. The refinery has spent $67 million since 1987 to reduce toxic emissions, which have dropped 74 percent since 1987.

A national study conducted in 1995 by the Environmental Defense Fund found the refinery to be among the best in the United States—and the second best in Texas—in terms of limiting pollution emissions per barrel of crude oil processed.

The refinery also received a prestigious national environmental award from Betz Laboratories, Inc. in 1994 in recognition of efforts to reduce the generation of solid waste and air emissions. The refinery has reduced sulfur dioxide emissions by more than fifty percent since 1997.

The refinery's parent company, ATOFINA Petrochemicals, Inc., has sales of $4 billion and employs about 2,000 people. The company is part of ATOFINA, the world's fifth largest chemical company, which, in turn, is the chemical division of TOTAL, which owns and operates twenty-eight refineries throughout the world and is a market leader in Europe and Africa.

The ATOFINA refinery at Port Arthur.
Sabine Lake can be seen in the background.

HOPEWELL OPERATING, INC.

Above: Bert Fields (in the driver's seat) working for Gilliland Oil Company in Burkburnett, Texas.

Below: Travis Hoffman (left) and Bert Fields (right) at Vance No. 1 Gas Well at Waskom, Texas, around 1946.

Hopewell Operating, Inc.'s roots began in the early 1920s with Bert Fields, Sr., son of Walter Smallwood and Berta Culberson Fields. Born and educated in Hillsboro, Texas, he left home at an early age. He initially purchased a flatbed truck and worked as an independent trucker. Bert entered the oil business as a contract driver using his own truck in the oil field during the oil play centered in Mexia, Texas, in the 1920s. The offices of Hopewell Operating, Inc. contain an undated panoramic photograph of the employees of Gilliland Oil Company in Burkburnett, Texas. Fields is seated behind the wheel of his flatbed truck, situated amongst the bunkhouses and derricks of a large developing oil field.

Bert Fields moved his family to Dallas in 1934. After entering the production side of the oil business, in the early 1930s, Fields added a drilling company to his expanding activities. By 1940, he was an oil operator, a producer, and a drilling contractor throughout Texas and even into Illinois.

His production included well-known Fields such as East Texas, SACROC, and the Osage Field in Oklahoma. He owned interests in the states of Alabama, Arkansas, Florida, Georgia, Illinois, Louisiana, Mississippi, New Mexico, Ohio, and Oklahoma, with the concentration of his assets in Texas.

Fields developed most of his properties without investors. His joint ownership was with major companies such as Conoco, Gulf, Hunt, Texas and Union Producing to name a few. He built his successful business on hard work, strong decisions, and good values.

In addition to being in the oil business, Fields spread his interests into ranching and banking. He was a member of the executive committee of the Texas Independent Producers and Royalty Owners Association. In the early 1960s he became the founding director of North Dallas Bank and Trust Company and named as chairman of the board in 1962. He was a member of the Dallas Petroleum Club and the East Texas Salt Water Disposal Board.

Bert Fields, Sr., died in 1963 leaving two children: a daughter, Jeanne Fields Shelby, and a son, Bert Fields, Jr.

Bert Fields, Jr., assumed operations and continued his father's oil and gas business. He also expanded the family's ranching, banking, and real estate holdings. The most notable banking interest being the North Dallas Bank and Trust (NDBT). After serving many years, Bert, Jr., still holds an active position on the board of directors at NDBT.

Bert, Jr., stepped down from the oil business to allow Bert Fields, Sr.'s grandsons to assume the family business. They have continued operations in Texas with Hopewell Operating, Inc., Shelby Resources, and New Waskom Gas Gathering.

Michael H. Shelby, Bert Fields' third generation, began his oil and gas business in 1971 while working with Bert Fields, Jr. By 1978 he was an independent producer in Texas under his own name and a corporation, Elgie Company, Inc., operating in Louisiana. In 1994, Michael merged all of his operations into Hopewell Operating, Inc. At the present time Hopewell is actively developing gas in the Cherokee Basin in Eastern Kansas. This development is a combined shale gas and coal bed methane gas play.

The family retains large interests in ranching and real estate. Throughout the state of Texas, they own and manage thousands of acres of timber, cattle, farmland, and wetlands. The family banking includes both North Dallas Bank and Trust and Legacy Bank of Texas. NDBT boasts being one of the largest independent banks in Dallas with the same name, same ownership, and profitability for every year of operation since its founding in 1961. NDBT has four banking centers located in North Texas. Bert Fields, Sr.'s younger grandson sits on the board of directors for Legacy Bank of Texas and serves as chairman of the board. This independent bank is a $600 million-plus, depository bank. Legacy has eleven branch offices located in Plano and surrounding cities.

As we move in to the twenty-first century we are seeing the family tradition continued with the fourth generation of Bert Fields' entering the oil business, real estate, ranching and banking with the same zest and enthusiasm as their great-grandfather.

All of this activity is the direct result of Bert Fields' initial foray into oil field contracting and has every indication of proceeding well into the current century, serving and growing in several industries and continuing his legacy into and beyond the fourth generation.

Left: Bert Fields.

Below: Michael H. Shelby at Novice, Texas, around 1993.

LEWIS B. BURLESON, INC.

Above: Lewis B. Burleson, Inc. employees at the Belding Yates Unit in Pecos, Texas (from left to right): Kathryn K. Burleson, Lewis B. Burleson, Nancy Burleson Egan, Cindy Blair, Sharon Beaver, Steven Burleson, Wayne Jarvis, and Julie Truitt.

Below: Lewis B. Burleson at Cook No. 2 in Lea County, New Mexico. This was the first producer for Burleson and Huff in 1967. This picture was taken in the fall of 2002, showing the current status of the well.

Lewis Burleson began his career in the oil industry as a geologist with Atlantic Refining Company in Texas, New Mexico and Louisiana for five years before becoming chief geologist for Joseph I. O'Neill, Jr. O'Neill was an independent oil producer in Midland often referred to as the "Scurry Reef King" because of his large holding in the Kelly Snyder Field. During this time, Burleson worked with a landman named Jack Huff and the two formed a good working relationship with O'Neill's company. Burleson and Huff quickly realized they could just as easily find oil for themselves. So they formed the Burleson and Huff partnership on February 1, 1959. On the day they opened their Midland office the Suez Canal crisis between Israel and Egypt began, throwing the region into tumult and causing a downturn in the oil business. The Texas Railroad Commission cut producing days to five per month, reducing the cash flow for oil producers and making it much harder to sell oil deals.

The company's first deal was the purchase and sale on the same day of an overriding royalty interest in New Mexico for a $1,000 profit. The company later sold its first drilling deal to Bond Oil Company of Dallas for a well in the Prentice Field of Terry County, Texas.

Between 1959 and 1968, the partnership put together forty drilling deals with the partnership keeping overrides and money as compensation. The partnership went into oil operations with the completion of the Cook No. 2 in the Jalmat Gas Field of Lea County,

New Mexico. Twenty more oil and gas wells were drilled in New Mexico including the Cooper No. 2, a shallow reef oil producer. The lease has made 280,000 barrels of oil to date.

Burleson and Huff eventually dissolved their partnership in May 1979 and Burleson chartered the family corporation of Lewis B. Burleson, Inc. Burleson took over the oil properties and Huff opened an office to pursue other oil ventures. Although Burleson altered his corporate structure, he remained true to the business approach of not getting involved in the drilling rig business or with allied companies in the service sector.

The years from 1982 to 1988 were a period of low gas takes and gas prices. In 1979, Congress passed the Natural Gas Policy Act to help gas production. The company's personnel studied and learned all about these new regulations. They used this knowledge to increase gas prices by knowing how to move a well from one category to another.

The company mandated several guidelines for drilling wells to allow it to excel and established a five-thousand-foot optimum depth to capitalize on the fact that the major portion of oil in the Permian Basin is shallow. The company also restricted the area in which it worked to counties west of Big Spring Street in Midland, an area that covered New Mexico and all the counties that contain the geological feature known as the Central Basin Platform.

Burleson's greatest accomplishment in the oil business came in the Belding Yates Field of Pecos County, Texas. He handled all the

geological, land, and engineering on the drilling of twenty-six oil wells in this field. The shallow Yates sand is being flooded with water and has produced as much oil in this secondary stage as it did in its primary stage and production continues to rise. This field has contributed greatly to the company's success and is expected to continue producing until 2025.

The company has three new production areas beginning in the late 1990s. A drilling program in the Fullerton Field of Andrews County, Texas, developed San Andres production. Twelve wells were drilled on this project. The company drilled additional wells in the Eumont Field of Lea County, New Mexico, increasing the company's gas reserves and Clearfork production on the Embar structure with the drilling of six wells on the company's Embar University leases. The company is now developing its Santa Rita lease, which is producing from the Abo section of the Permian. Additional wells will be drilled in the next two years.

A production curve for the history of the company shows high gas volumes averaging more than 3,000 mcf per day with oil production at 700 barrels per day in 2003. A similar increase in oil and gas production is expected in the coming years. The curve has been increasing with time.

Today, Lewis B. Burleson, Inc., is a Midland, Texas, oil company with a production yard in Jal, New Mexico. The company began with two people and expanded to three with a secretary and bookkeeper. In 1979, the year the company formed, it hired a field superintendent to handle the company's daily work and three pumpers.

Ten years later the company hired a bookkeeper to deal with the huge amount of paperwork the business generated. In 1983 Steven Burleson, son of the company founder, was hired as vice president to handle geological and field operations after earning a degree in geology from Texas Tech University.

Steven Burleson now makes the bulk of the decisions at the company and will rise to president when his father retires. The company name will change to Burleson Petroleum to reflect the new ownership, while the approach to the oil business will remain the same.

Above: The Production Facility Cooper No. 2 Well Shallow Reef Producer Lea County, New Mexico, the company's best productive lease.

Below: A graph of Lewis. B. Burleson, Inc.'s Permian Basin Production from 1958 through 2003.

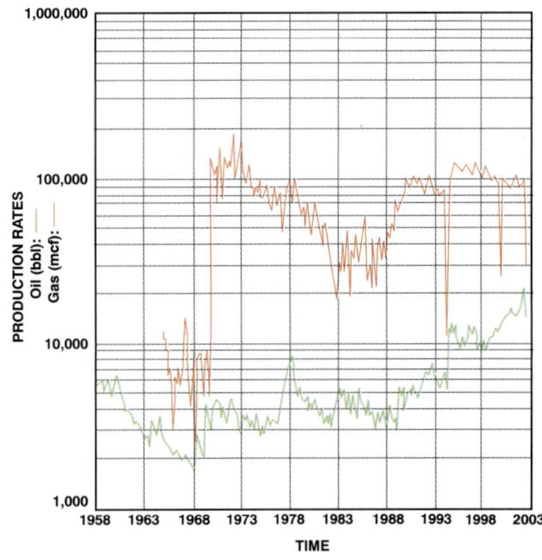

C&K
Petroleum

Above: W. D. Kennedy and
C. Fred Chambers.

Below: W. D. Kennedy on a job site with an
employee.

C&K Petroleum was founded in 1953 from a Midland partnership between C. Fred Chambers and W. D. Kennedy, both natives of Dallas. The men drilled two dry holes that year and after drilling four more dry wildcat wells a well was spudded in Ector County and another in Tom Green County.

Chambers and Kennedy were trying to decide whether the partnership should be called Chambers and Kennedy or Kennedy and Chambers. To settle the issue, they decided each would drill a well and if either produced, the winner would be entitled to have his name first in the partnership.

Both wells came in. But the one Chambers selected was completed first and Chambers and Kennedy became the official name. In 1959 the partnership expanded with the opening of an exploration office in Houston. Houston then became the company's headquarters and as C&K expanded the company opened operating offices in Denver, New Orleans, Jackson, and Lafayette.

In 1966 the partnership formed a Canadian corporation with J. M. C. Ritchie of New York. The company, called Ritchie Oil C&K (later known as the ROCK), had its headquarters in Alberta, Canada.

The company expanded further by forming a corporation, eventually known as C&K Nederland Company, to acquire a concession from the Dutch government. On August 31, 1970, the partners exchanged their assets for C&K Petroleum, Inc., common stock.

On that day, C&K acquired all of the assets of Nortex Oil & Gas Corporation, one hundred percent of the stock of Valley Royalty Corporation, and certain production payments of Southwestern Life Insurance Company. The company went public in August of 1971 and was listed on the American Stock Exchange in November 1972.

Operating results for that year reflected the company's success. Net income for the year rose eighteen percent over those recorded the preceding year due largely to the company's experience and proficiency as specialists in the search for and production of oil and gas.

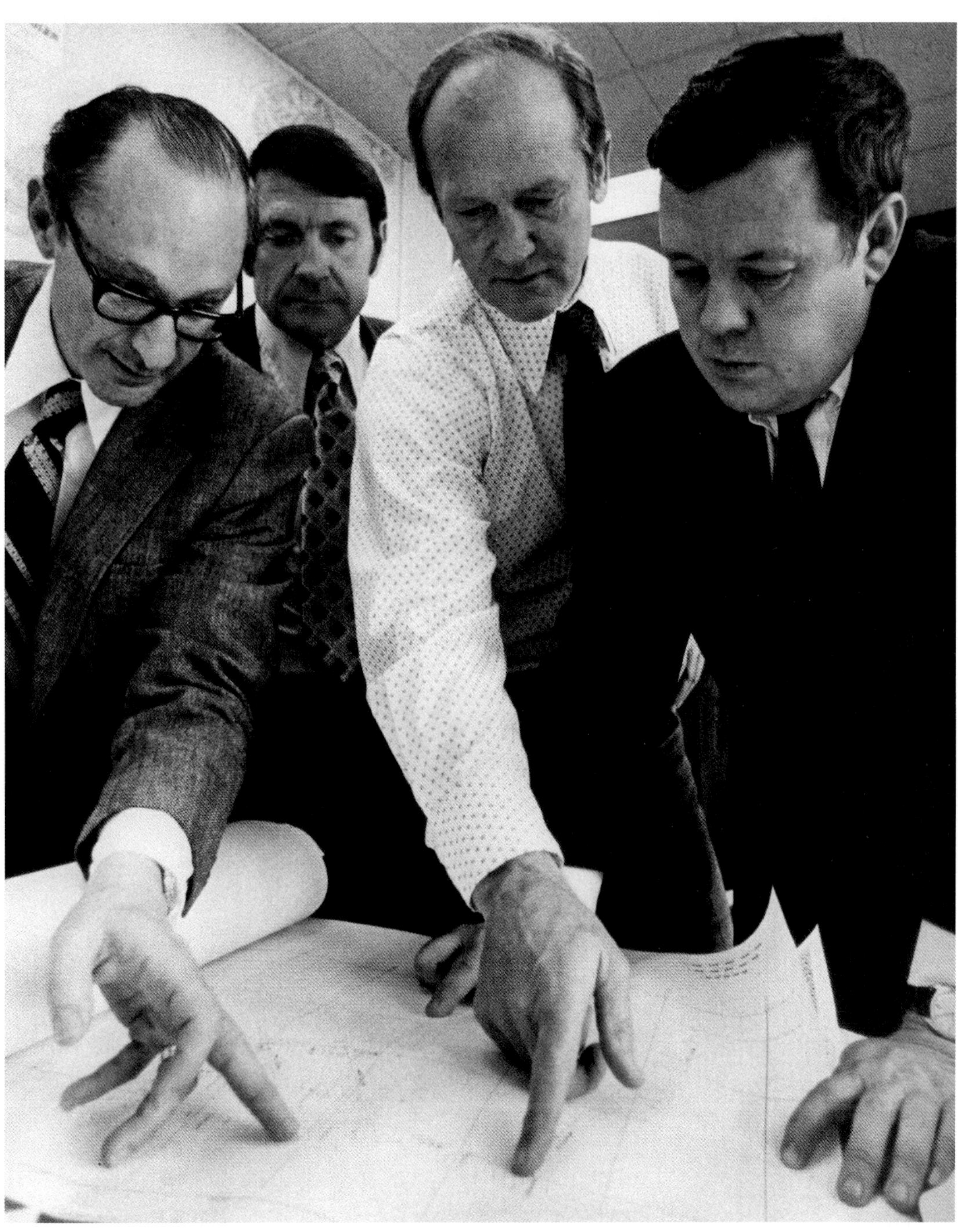

C. Fred Chambers in a planning meeting.

The majority of C&K's oil reserves were in Texas and Mississippi, with gas reserves primarily in Louisiana and Texas. The company also explored reserves in other parts of the world including the McKenzie Delta, Beauford Sea, and the Arctic Islands.

Chambers was a longtime friend of former President George W. Bush, who served as an honorary pallbearer at his funeral while president in 1989. Chambers helped raise funds for Bush early in his political career.

President Bush and Chambers met in the early 1950s when the two worked in the oil business in Midland. Chambers served as campaign manager for Bush's unsuccessful 1970 run for the U.S. Senate and for his successful 1966 congressional campaign. He served as a longtime fundraiser for the Bush campaigns. The two also had joint business dealings.

Chambers and Kennedy eventually went on to other things when they sold the company to Alaska Interstate in 1980. Kennedy died in 2002.

Kennedy graduated from the University of Texas with a business degree and was a member of the Independent Petroleum Association of America and the American Wildcatters Association.

Thomas D. "Rusty" Howell

"A true oil man is someone who will gamble his own money on a hole in the ground," says Thomas D. "Rusty" Howell of Marshall, Texas.

Howell has seen his share of "holes in the ground"—a euphemism for oil and natural gas wells. Most have been productive because Howell is a student of petroleum exploration, discovery, and production. He remains one of the most knowledgeable men in the industry about how and where to coax the mineral of increasing value from its hiding places in the earth. He also is regarded as a superior troubleshooter when something goes wrong in that "hole in the ground."

Born in Comanche, Texas, on June 28, 1928, to Lamar and Opal Howell, he excelled in school, graduating from high school two years ahead of schedule. After graduating from Texas A&M in 1948, Rusty worked briefly in Raymondville before accepting a position as assistant county agent in Marshall, Texas, and often working in nearby oil fields by night. He married Joy Lois Etheridge in 1951, the same year he went to work for Halliburton, an oil field service company, as a cement truck driver. In 1953, Rusty joined Tri-State Oil Tools, Inc. During his years with Tri-State, he managed Tri-State operations in Kilgore, Palestine, and Beaumont in Texas; Morgan City in Louisiana; and the northeast region of the United States while based in

Above: Thomas D. "Rusty" Howell.

Below: Rusty and Joy Howell.

Pittsburgh, Pennsylvania. He had become vice president of international operations for Tri-State when he left the company twenty-five years later.

"I wanted to stay in the field," says Rusty, "and learn all I could about what went on down in those holes." Howell became an expert on how to solve problems with troubled wells.

In 1978, Rusty became a co-owner and vice president of Carlile and Howell, Inc.; Marshall Exploration, Inc.; Martex Drilling Company; and H&C Well Service. Since 1990 he has continued to develop drilling prospects and invest in oil and gas exploration and development activities in association with Howell Oil & Gas, Inc., an ongoing oil and gas business owned and operated by son Steve and daughter Leslie.

Rusty has served his community and industry in various capacities. At home, Howell was instrumental in the formation of the Marshall Economic Development Council and served twice as chairman. For the oil and gas industry, he served multiple terms as president of the Texas Independent Producers and Royalty Owners Association, and was named "Mr. TIPRO" for 2001 because of his effectiveness in representing the organization and the industry to state

and federal officials. He is also a charter member of the Texas Rangers Association Foundation, an organization formed to support and preserve the Texas Rangers, Texas' elite law enforcement officers.

Despite—or because of—his outstanding success in the oil and gas business, Rusty is most likely to be found in his office in Marshall dressed in khakis and a plaid shirt, keeping an eye on various aspects of his continuing business efforts. He is also quick to credit wife Joy with their success: "She always encouraged me, and I missed many a Christmas when we had some problem in the oil field, but she understood."

Rusty's ability to understand all aspects of the oil and gas industry has been key to his success. He learned how to deal with oil field problems literally by dealing with them and remembering what worked and what did not work. When nothing worked, Rusty would often invent a tool or procedure that would, as evidenced by several patents he holds for downhole tools. He believes that success comes from regarding the oil business as a business, period, and not glamorizing it as a get-rich-quick game. "Learn to evaluate," he says. Above all, he recommends studying the cycles of the industry to be

in position to gain on each inevitable rise and fall.

"Rusty" Howell has been around the industry long enough, and is successful enough, to have earned recognition as an "original—the best there is at what he does." That accolade would likely produce a disbelieving smile and a twinkle in his eye. Rusty has been doing it now for half a century.

Above: Rusty Howell with his longtime friend, former Speaker of the House of Representatives Pete Laney.

Below: The "Mr. TIPRO" Award is presented to Rusty Howell by TIPRO President Rex H. White, Jr., in recognition of Howell's service to the organization.

SOUTHEASTERN PIPE LINE COMPANY

Southeastern Pipe Line Company was incorporated in 1960 by Stanley C. Woods to transport gas from Appling Field in Calhoun County, Texas to the Tennessee Gas Pipeline eight-miles away. At its peak the Southeastern line moved approximately twenty-two million cubic feet of gas daily. During the early days of Woods' company, Richard Triche, John Stoker, L. P. Jones, and C. A. Reed made significant contributions to Southeastern's operations.

Shortly after incorporation, the Aluminum Company of America (Alcoa) and Crown Central Petroleum began attempting to block the transporting of gas from Appling Field to market, and to shut down wells drilled on small tract lots. In 1962, Stanley Woods, Southeastern and Woods Exploration & Producing Company sued Alcoa and Crown for violating state and federal antitrust laws. They would spend the next thirteen years in legal battles for survival.

After a six-week trial in U.S. District Court in Houston in 1968, a jury found Alcoa and Crown in violation of the Sherman Antitrust Act by conspiring to monopolize, attempting to monopolize, and monopolizing the Appling Field. The presiding judge set aside the verdict, however, and granted judgment in favor of Alcoa and Crown (*Woods Exploration & Producing Co. v. Aluminum Co. of America,* 304 F.Supp. 845, S.D. Tex. 1969). The appeals court directed the lower court, to reinstate the jury verdict for Southeastern, and ordered trial of a portion of the case the lower court had

Above: Stanley C. Woods.

Below: East Bernard Field in Wharton County, Texas.

dismissed before trial (*Woods Exploration & Producing Co. v. Aluminum Co. of America,* 438 F.2d 1286, 5th Cir. 1971). The U.S. Supreme Court refused to hear further appeal from Alcoa and Crown.

After a six-week trial in 1972, a decade after litigation began, a second jury found for Southeastern. Alcoa and Crown appealed, but the Fifth Circuit Court upheld the judgment for Southeastern and Woods (*Woods Exploration & Producing Co. v. Aluminum Co. of America,* 509 F.2d 784, 5th Cir. 1975). In 1975 the U.S. Supreme Court refused to hear Alcoa's and Crown's last appeal.

Southeastern's other battles with Alcoa and Crown created further changes in Texas law. In 1963 the Texas Supreme Court held that each producing zone was a separate field for the purpose of apportioning monthly production allowables (*Benz-Stoddard v. Aluminum Co. of America,* 368 S.W.2d 94, Tex. 1963). A year later the Texas Supreme Court again ruled for Southeastern in *Railroad Commission of Texas v.*

Aluminum Co. of America (380 S.W.2d 599, Tex. 1964). This upheld the Railroad Commission's decision to reject Alcoa's application to change the field rules setting production allowables for the Appling field, based one-third on well and two-thirds on acreage. The court held that laches barred Alcoa's efforts to change the rules for setting allowables in established fields where producers had relied on existing rules to invest substantial money drilling wells. This holding affected approximately twenty-five thousand wells drilled in old, established fields, especially the East Texas and Yates Fields. The rule Alcoa sought would have had a devastating effect on Texas landowners and small operators.

Three months later, the Texas Court of Civil Appeals held that Southeastern could sue Alcoa and Crown in state court for violating Texas antitrust laws, while also suing in federal court for violating federal antitrust laws (*Woods Exploration & Producing Co. v. Aluminum Co. of America*, 382 S.W.2d 343, Tex. Civ. App.–Corpus Christi 1964). The Texas Supreme Court refused to hear further appeal of that case.

In 1966 the Texas Supreme Court again ruled for Southeastern in *Railroad Commission of Texas v. Woods Exploration & Producing Co.* (405 S.W.2d 313, Tex. 1966). The court upheld the market-demand rule when Southeastern challenged the Commission's power to grant Alcoa's request to set production allowables for the Appling Field at less than the market demand.

Alcoa's and Crown's actions and the seemingly endless litigation drained Southeastern's financial resources, leaving it facing bankruptcy at one point. In another wily legal move, Southeastern agreed to a receivership, allowing its continued operation until it finally prevailed in its litigation.

During the 1970s, Southeastern changed its corporate charter from a pipeline to a producing company, drilling wells to explore for and produce oil and gas in fields including East Bernard and Orchard.

Stanley Woods has been a force throughout the oil industry. He was instrumental in forming the Texas Landowners and Independent Oil and Gas Producers Association, composed of sixty-five hundred landowners and small independent producers. This group successfully lobbied against efforts to create force unitization in Texas. Woods has also been a member of the Texas Independent and Royalty Owners Association since its early days. He testified before the U.S. Senate Finance Committee to support a 27.5 percent depletion allowance, and worked for an import tax on foreign oil to encourage U.S. exploration. Through Woods Exploration & Producing Company, he has drilled Texas wells in the Appling, Diebel, Fannett, and Magnolia Beach Fields, and in Louisiana, Alabama, and Florida.

Woods remains as Southeastern's president. Other officers are Ann Woods Bozeman, vice president; and Louise Geno, corporate secretary. Consultants include David and Gary Bagnall, engineers; Robert Fendley, geologist; Harley Jones, pumper in the East Bernard Field office; and Roy Barnes, attorney at law for Southeastern.

From 1959 until October 1975, th late Levert Able successfully represented Stanley C. Woods and Southeastern Pipe Line Company et al in its many years of antitrust litigation against Aluminum Company of America et al.

Southeastern's headquarters is at 806 Main Street, Suite 910, Houston, Texas 77002.

Stanley C. Woods incorporated Southeastern Pipe Line in 1960.

M. B. "Duke" Rudman, one of the all-time, most successful independent oilmen and an original East Texas wildcatter, is also among the most colorful men in the oil industry. During his more than sixty years in the business, Rudman has drilled thousands of wildcat wells throughout the United States, from North Dakota to Texas, and from California to Florida.

A legend for his personal sense of style, his dedication to fitness and health, and his philanthropic endeavors, Rudman is the embodiment of the classic American success story. A native of Bonham, Texas, he grew up in humble surroundings. "I sold newspapers and shined shoes," he recalls. "When I made a dime, it was a big day. Lots of days there was only a tin of sardines to eat."

As a Jewish boy growing up in an Oklahoma mining town, he was regularly taunted and pummeled by other kids—until a neighbor who had boxed at Notre Dame taught him how to fight. The abuse from his peers stopped abruptly.

Even as a boy he was determined to become somebody, and says his serious differences with his father probably afforded him something of the challenge he needed.

Rudman became interested in the oil business at a young age, after admiring several oilmen with whom he came in contact. After attending Kemper Military Academy and the University of Oklahoma he began hustling oil deals, living in a pipe yard to make ends meet. He studied every phase of the oil business, moving to the East Texas oil fields during the

boom of 1931. He carved a successful niche for himself by taking over leases that had been dry holes and turning them into productive wells, and also helped develop several technical innovations for the oil industry. By age thirty he had interests in 175 wells and an income of $5,000 a month.

He explored acreage in North Dakota's Williston Basin that became one of the major fields of that era. Rudman's accomplishments have continued, resulting in an overall success ratio that today exceeds ninety percent of all other wildcatters, past and present.

He moved to Dallas in 1942 and now spends most of his time in his office there, waiting for deals to come to him. He generally takes about twenty-five percent interest in a well and looks for those with the best potential to produce a monster field. Although this presents a greater risk of a dry hole, it also offers the possibility of hitting the big one, a gamble that Rudman relishes. "That's what drives me," he says. "That's what this business is all about."

One key to his success has been the timeliness of his business decisions. In June 1981, he foresaw that the debt load of several oil-producing countries would lead to what became the U.S. energy crisis. He responded to this realization by selling ninety-five percent of his oil and gas reserves at the peak of the oil boom, just prior to the dramatic decline in the price of oil. His ventures since that time have been financed with the interest from his investment of those profits in Treasury bills.

Described by a co-worker as "a generous, empathetic human being," Rudman has shared his wealth through philanthropy to a wide variety of causes. These range from Jewish charities to the Greater Dallas Crime Commission, and from the Dallas Children's Theater to the Crystal Ball Charity.

Now in his nineties, Rudman continues to maintain an office in downtown Dallas as the principal of Rudman Partnership. He lives by the principles of honesty and integrity. "Be true to your word," he says. "Your word is your bond." This is just one of the "rules" at the foundation of his business success.

He also advises that those who want to get ahead in business know their product, find a mentor who has made it in that business, and then share their own knowledge with others. He advocates hard work, noting, "The harder you work, the luckier you'll be."

"Persistence in business is better than graduating summa cum laude," Rudman says, adding that his own experience bears this out. He drilled twenty-nine straight dry holes before he hit a good one. His other credos include practicing self-denial, believing in yourself, setting your sights high, and being courteous to everyone.

And a final rule—one for which he has won awards—is to look good at all times. Rudman's adherence to that philosophy has placed the Dallas oilman at the top of the list of "Best-Dressed Men in Europe." Scores of newspaper and magazine photographs over the years have pictured him wearing everything from mod fashions to spats, a Victorian-style frockcoat to an unusual snap-brim bowler hat and velvet vests.

SUEMAUR CORPORATION AND SUEMAUR EXPLORATION, INC.

William B. Miller and Daniel A. Pedrotti, both experienced South Texas exploration geologists with major company backgrounds, operated for several years as Miller & Pedrotti before forming Suemaur Corporation on June 4, 1975.

The discovery of the Southeast Charmousca Wilcox Field by the Miller-Pedrotti No. 1 288-DCRC in Duval County on December 20, 1973, provided the capital needed to participate with Frank Morrison, Leo McClosky, and Gordon Lansford in the purchase of three drilling rigs from Baggett Drilling Company. This led to the founding of Cinco Drilling Company, which ultimately became a successful five-rig company that was sold to Delhi International in 1978.

With the formation of Suemaur Corporation, Miller and Pedrotti defined drillable prospects in the "On-Shore Province" of the South Texas Gulf Coast. They then assembled acreage blocks on these prospects and found drilling partners to fund evaluation of the oil and gas potential.

Suemaur Corporation continued to pursue drilling deals in the non-pressured Jackson-Yegua-Queen City-Wilcox Trends as well as the Frio-Vicksburg Trends of Texas Railroad Commission Districts 1, 2, and 4. In the late 1970s, Suemaur acquired a five-thousand-acre lease block in the Lobo Trend that Getty Oil Company drilled, leading to the discovery of the El Gato Field in 1980. This field, in which Miller and Pedrotti retained overrides, has produced more than 80 billion cubic feet of gas, 650,000 barrels of condensate and is still active.

The partners soon realized the importance of controlling the operations of their wells and on April 3, 1984, founded Suemaur Exploration, Inc., turning over exploration, development, and operational activities to the new company.

Dovie Price, the company's longest full-time employee, joined the company on July 6, 1976, as a receptionist-bookkeeper and today is the secretary-controller of Suemaur Corporation. Rosanne Claunch served as secretary-controller of Suemaur Exploration, Inc.

Buying the exploration assets of Harkins Drilling Company in 1990 and hiring its exploration staff required a change in business philosophy. Prior to this purchase, Suemaur confined its drilling activities to the non-pressured prospective sections of the Frio-Wilcox and those not requiring fracture treatments.

The Harkins exploration prospects were focused on the reservoirs below the first string of protection pipe and many would require fracture treatments. This launched a new era for the company. Using the revenue stream from the recently discovered Suemaur Field, Suemaur Exploration embarked on the pursuit of deeper and larger potential reserves by joining more permanent industry partners.

Suemaur Corporation and Suemaur Exploration, Inc. continues to do business today. But the major exploration and production activities are handled through Suemaur Exploration & Production, LLC.

Suemaur Exploration & Production, LLC, is a Corpus Christi company that developed as an offshoot of Suemaur Corporation in 1990. The men behind the company—William D. Miller and Daniel A. Pedrotti—worked for years as partners before officially incorporating as Suemaur Corporation in 1975.

The two men formed Suemaur Exploration, Inc., because they understood the value in controlling their wells. By forming the company, they were able to turn over the exploration, development and operational activities.

In January 1990, Miller, Pedrotti, and J. M. Smith bought the exploration assets of Harkins & Company and assigned them to Suemaur Exploration, Inc. One outgrowth of this new entity was the formation of various limited partnerships to hold the real property assets generated by the company.

Suemaur Exploration functioned as the general partner for the limited partnerships and served as the operator for all E&P operations until the end of 1997. In October 1997, Moorehead Oil & Gas became an equity owner and a new partnership was formed—Suemaur Exploration Partners.

Suemaur Exploration & Production, LLC, which was formed in January 1998, manage the producing assets and the exploration activity of the partnership. In July 1998 Lucien Flournoy replaced J. M. Smith as a partner and equity owner in Suemaur Exploration Partners.

Since 1970 the various Suemaur companies have originated and participated in more than forty new field discoveries along the Texas Gulf Coast. Suemaur Exploration & Production, LLC, functions as an exploration cooperative.

In addition to Suemaur, there are three outside industry partners that comprise the venture. The outside partners are Hurd Enterprises, Ltd., FESCO, Ltd., and CL&F Resources, LP. All exploration expenditures, including overhead, seismic, land, drilling and completion costs are borne in proportion to each company's interest in the joint venture and the prospects.

Traditionally, independent exploration companies have maintained a staff and database that is large enough to generate and drill a sufficient number of quality prospects to assure statistical and economic success. In most cases, independents do not retain one hundred percent interest in their internally generated prospects and are constantly seeking partners, a process that is time consuming and expensive.

The exploration cooperative concept allows participating companies to share exploration costs rather than duplicating them. Each partner has the option of retaining a fixed working interest in any or all of the generated prospects. At present, Suemaur Exploration & Production, LLC, is staffed to handle all phases of exploration, drilling, marketing, production, and divestiture.

The company leader is President Robert W. Maxwell. The company employs around thirty people and one consultant representing a number of professional disciplines, including geology, geophysics, land, engineering, and accounting. The company's exploration staff is formed into teams consisting of two geologists, a geophysicist, a geophysical workstation, and a geotech. Together, these teams explore along the Texas Gulf Coast from the Rio Grande to the Sabine River.

SUEMAUR EXPLORATION & PRODUCTION, LLC

EASTLAND EXPLORATION, INC.

Eastland Exploration, Inc., traces its beginning to A. M. Donnelly, who, in the early 1900s, developed the natural gas resources and distribution system for the City of Buffalo, New York, and Windsor, Ontario.

In 1909, Donnelly moved to Tulsa, Oklahoma and operated in the Nowata and Bartlesville area, drilling wells as a drilling contractor and for his own account. He established the A. M. Donnelly Drilling Company on March 4, 1916.

He later moved his operation to the Eastland, Texas area where on January 22, 1922; he founded The Eastland Oil Company. By this time, he had moved to Fort Worth, which became the headquarters of the company.

During this period he discovered the Martin Pool in Eastland County and the Eastland Pool in Coleman County. A. M. Donnelly died in June 1927 and the presidency passed to George A. Donnelly whose sons, George A. Donnelly, Jr., and Richard Donnelly, would later play key roles in the company.

Eastland was an early pioneer in the Permian Basin of West Texas and New Mexico. In 1926, Eastland was the drilling contractor on the R. A. Westbrook, T. G. Hendricks No. 1A, the discovery well of the great Hendricks Field and the first oil well discovered in Winkler County.

In the Permian Basin, Eastland was responsible for the discovery of the Donnelly (San Andres) Field, Donnelly (Cisco) Field, and the Darmer (Queen) and Power (Grayburg) Fields. The company was instrumental in the discovery of the Cowden (Cisco), Darmer (Canyon) and the Champion Lake Fields.

The company developed the TXL (Penn), Sable San Andres, and Herradura Bend Fields. Production discoveries range in depth from 330 feet in Sandoval County, New Mexico to 13,020 feet in Eddy County, New Mexico.

In 1956, George A. Donnelly created the Eastland Drilling Company and operated it and The Eastland Oil Company as separate entities. On August 1, 1959, The Eastland Oil Company sold its assets to Sinclair Oil Corporation.

On January 1, 1970, the drilling company changed its name back to The Eastland Oil Company. Donnelly died on August 22, 1976, and his oldest son, George A. Donnelly, Jr., became president with Richard Donnelly serving as vice president.

In 1984 the oil company created two subsidiaries and, on January 1, 1988, Eastland Exploration, Inc., became a separate company owned by Richard Donnelly, a grandson of the original founder of The Eastland Oil Company. He now serves as president and Thomas A. Sikes serves as vice president.

Eastland Exploration, Inc., continues to participate in the oil and gas business throughout the United States. Project participation by the company and its principals has ranged from the North Slope to Colombia and from California to Illinois.

Above: Eastland Oil Company Mason No. 1, Ward County..

Generation to generation, The Eastland Oil Company continues to be a consistent participant in the oil industry. From its beginning with Arthur Matthew (A. M.) Donnelly, the company has been a stalwart in exploration and production of oil and gas in Texas.

Donnelly started his career in the gas fields near Windsor, Ontario, and Buffalo, New York, around 1891. George A. Donnelly, his son, working for A. M. Donnelly, drilled his first well in Texas in Denton County in 1914.

The Eastland Oil Company was incorporated in Texas in 1922 to produce oil from the Martin Field discovery. In those early years, Eastland drilled wells throughout the Eastland, Ranger, and Cisco areas of Texas.

In 1926, Eastland was the drilling contractor on Roy West's T. G. Hendricks No. 2, the first oil discovery in Winkler County, Texas.

The Eastland Oil Company and George A. Donnelly drilled in some of the biggest fields in the state. He drilled in the Penwell, South Ward, McCamey, Hendricks, Kermit, Hobbs, Westbrook, and Wasson Fields.

George A. Donnelly, Jr., joined Eastland again in 1946 after a tour of duty with the U.S. Army Air Corps. Receiving his degree in geology from the University of Texas in 1940, he was instrumental in Eastland's developing the Donnelly (Cisco) Field, the Donnelly (San Andres) Field, Darmer (Queen) Field, the TXL (Pennsylvanian) Field, the Sable (San Andres) Field, the Power (Grayburg) Field, and the Champion Lake Field.

THE EASTLAND OIL COMPANY

In 1978, George, Jr., became president and the headquarters of Eastland was moved to Midland. Richard Donnelly, George, Jr.'s brother, joined the company in 1949 after a tour of duty with the Navy and became vice president of Eastland in 1977.

Exploration activities ranged from the Black Warrior Basin to the Williston Basin to Southeastern New Mexico to Colombia in South America.

In July 1983, Robert R. "Robin" Donnelly, the fourth generation of the Donnelly oil men, joined Eastland, having worked internationally and then with a series of independent companies in the United States. Now, Robin serves as president of Eastland.

Currently, Eastland operates leases in the Pecos Slope area, the Artesia, Turkey Track, Parklane, and West Harper Fields, and in Starr County, Texas.

As the millennium begins, The Eastland Oil Company continues in its role as an independent oil and gas producer with emphasis on exploration and production of oil and gas.

Above: Frank Hatton finishing a bit at Boyer Well in 1957.

Below: Eastland Oil Rig No. 2.

HUDDLESTON & CO., INC.

Left: B. P. Huddleston.

Right: Peter D. Huddleston.

Founded by B. P. Huddleston, P.E. in 1967, Huddleston & Co., Inc., has enjoyed continuous founder and successor management. Now serving as chairman of the board, B. P. Huddleston remains active in the firm, participating in daily management and performing engineering studies. He also founded Peter Paul Petroleum Company in 1973 and serves as its chairman. His fifty years of experience in the petroleum industry began with summer work in West Texas oil fields as a teenager. Huddleston received his bachelor of science degree in petroleum engineering in 1957 from Texas A&M University, where he was a "Distinguished Military Student" and 1955 captain of the varsity football team. He served from 1981 to 1998 as a visiting professor of engineering at A&M, and has received numerous professional honors and awards.

Huddleston & Co., which has been led by Peter D. Huddleston as president since 1989, provides petroleum engineering consulting services for oil and gas and gas transmission companies, banks, and other financial institutions. The company provides business and financial services relating to virtually all phases of exploration and production activities, as well as technical services. Its client base includes entities of over five hundred major oil companies, large and small independent companies, gas transmission companies, financial institutions, government agencies, law firms, and banks, in North America and internationally.

With a database of over one hundred thousand entries, Huddleston & Co. has provided reservoir evaluation for virtually all-producing basins in the United States and Canada. International technical projects as well as individual exploration and exploitation projects have been performed in virtually all parts of the world.

Huddleston also provides gas transportation studies for the design and upgrading of facilities, pipelines, and gathering systems. These studies include evaluating existing gas supplies, potential future discoveries and extensions, all on a property-by-property and total service area basis.

The company evaluates the economics of exploration and production, covering estimated reserves and future production/price schedules and cost projections of market values and other investment parameters to determine investment goals and risk management. Huddleston's advice to financial institutions, pension managers, and insurance companies since 1978 has involved direct investments of approximately $2 billion.

Huddleston & Co. continually reviews technological advances in reservoir engineering with consideration for the company's historical evaluation techniques, to utilize improvements in engineering technology, computer processing capabilities, and economic advances relating to petroleum evaluation.

For more information on Huddleston & Co., Inc., please visit www.huddlestonco.com.

A privately owned oil and gas exploration company, Tri-C Resources, Inc., was formed in 1985 to pursue exploration and development opportunities on the Texas and Louisiana Gulf Coast.

The company began when two young Texas Oil & Gas Corporation employees, Scott Cone, an engineer, and John U. Clinch, a geologist, wanted to start their own company. Two investors, Michael M. Cone and David Carmichael, joined them by putting up the seed money to found the company, which they named Tri-C Resources, Inc. Together with Bo Herrin, a landman, these founders were instrumental in developing Tri-C during its early years.

They began by putting together leases offsetting production with geological control, and drilling shallow wells, which ranged from 3,500 to 6,500 feet in depth.

Tri-C, which has seventeen employees, functions on an annual budget of $20 to $25 million expended on a portfolio of high quality, diverse prospects. It operates a majority of the wells to contain costs, exposing risk capital to drill prospects and utilizing short term, multi-well commitments and farm-out opportunities to decrease the cycle time of prospect to pipeline.

The company has drilled about 600 wells in the past seventeen years, with a commercial success rate of approximately 70 percent. This ratio has been maintained, even though well depth has more than doubled in that time—from 5,000 feet in 1985 to a current average of 12,000 feet.

Tri-C's success in getting wells drilled and finding production with outside investors led within a couple of years of the company's formation to establishment of a G&A program that has resulted in a group of investors who have remained with Tri-C for some fifteen years.

This private investment group, along with Tri-C employees and principals, pays for or finances the company's exploration expenses and participates in its projects today, as they have since the company's formation.

Originally focused on the low-risk, low-cost development arena, where profitability is determined by efficiency, Tri-C's exploration strategy now includes a broader prospect portfolio in terms of both risk and potential. Despite the broadening of its development scope, Tri-C Resources continues to take pride in maintaining its tradition as a low-cost operator at greater drilling depths. The company's dedication to this practice is reinforced by the fact that Tri-C insiders own fifty-five percent of the company's exploration program.

For more information about Tri-C, which has offices located at 909 Wirt Road in Houston, please call 713-685-3600 or fax 713-685-3630. The company's e-mail address is info@tricresources.com and its website is www.tricresources.com.

Above: (From left to right): Scott Cone, John Clinch and Jim Cisarik on rig.

Below: (From left to right): Mike Cone, Ed Austin, David Carmichael, Scott Cone, and John Clinch.

Hy-Bon Engineering

World War II veteran Ralph Nelson, who saw the need to remove dangerous gas emissions from the air, founded Hy-Bon Engineering in 1952 in a West Texas garage. During that time, the West Texas sky was ablaze with natural gas flares, and much of the remaining gas was simply vented to the atmosphere. Beyond the obvious health problems this gas and smoke caused, several deaths had occurred due to H2S poisoning. Nelson identified the need for a way to capture these gases economically. After experimenting with different metals that were immune to the corrosive powers of H2S, he developed the first vapor recovery unit, coining the term.

Hy-Bon has grown from that basic vapor recovery unit design to become one of the most well respected companies in the industry specializing in gas management systems. The company focuses on a particular niche of the market—handling low-pressure gas—and is a global leader in this arena. Capturing very low-pressure gas is a challenging engineering feat, and one that has taken the company around the globe and given it the opportunity to assist several industries.

Jim Woodcock, who acquired the company in 1981 and serves as chairman of the board, expanded the company to pursue methane gas escaping from any industry—ranging from landfills to breweries. For example, Hy-Bon equipment is used to operate Mexican breweries by capturing the fumes off fermenting beer and converting it into energy that runs the plants via gas driven generators. A Hy-Bon unit successfully draws methane gas from a wastewater treatment operation in New York, which is then used to power gas generators, providing electricity for an entire neighborhood. In a peach-canning factory in California, the peelings are made into sludge, heated, and bacteria are added to emit methane, which is then captured by Hy-Bon equipment and used to power the cannery.

However, Hy-Bon did not abandon its roots. The company has also expanded its role in the oil and gas industry beyond vapor recovery to include gas booster systems, casing head pressure reducing units, and a range of compressor packages focused on handling low-pressure gas. Hy-Bon has been fortunate to provide units for the oil industry in over twenty countries, both onshore and offshore, in an incredible range of applications.

"A lot of companies sell products, at Hy-Bon we work with our customers to provide solutions," Hy-Bon President Larry Richards says. "In most instances, we can eliminate an environmental problem and simultaneously make our customers money from a gas stream they already own—a rare combination!

"Hy-Bon Engineering has been honored to serve our customers over the past 50 years and may the next 50 be equally as exciting!"

Hy-Bon Engineering is located at 2404 Commerce in Midland, Texas and on the Internet at www.hy-bon.com.

Malone Petroleum Consulting & Joint Interest Auditors recovers millions of dollars for its clients. Its two largest client recoveries—$60 million and $18 million—had already been audited and missed. It has had over one thousand findings in some audit reports, and has even obtained a credit for an entire well's drilling charges for a client. An interesting recovery was $5.2 million—but it only reviewed $2.4 million in cost. The company performs beneficial audits of acquisitions, blowouts, contracts, expenses, legal, non-consents, payouts, and revenues. The owner is Robert P. Malone, who was trained by a major oil company in the oil field and who is known as the "auditor's auditor." He is a Council of Petroleum Accountants Societies expert and is considered one of the best auditors in the world.

Externally, Joint Interest Auditors is a proven, valuable profit and information center for management. It has always been a wise policy to audit investments in order to improve profitability by recovering and precluding incorrect overcharges from outside operators.

Internally, Malone Petroleum Consulting provides services, either on a day rate and/or contingency basis, in order to reduce client losses that happen in every organization resulting in significant, cumulative losses. MPC has learned from the best in the industry.

The organization would not begin to be what it is today without the genius and support of certain individuals—Jim Boney, Dexter Earl Brown, Barry Cohen, Robert A. Giroir, Olga Dubois Malone, Daniel A. Nelis, Lee A. Perry, Leigh Prieto, and Stephen Tredennick.

Malone has a favorite story: ARCO was a very big client and Malone had never met J. R. Ashley, who was ARCO's director of joint interest audit. So Malone called Ashley and asked to meet

him. Ashley said, "I don't want to meet you, Bob, because in my mind, you leap tall buildings! If I meet you, I will see that you are fat and ugly and it will destroy my image of you." Malone said, "I think you are serious." Ashley said, "Yes, I am very serious." (Malone and Ashley did meet much later.)

Malone, who is a certified fraud examiner, worked directly with Raymond Plank, president of Apache. The audit hit the front page of the *Wall Street Journal* and Apache's stock went up after the article was published.

MPC is located at 12651 Briar Forest #165, Houston, Texas 77077. For more information, please call 281-531-1500 or visit wwwmalonepetroleumconsulting.com.

CLAYTON W. WILLIAMS

Clayton W. Williams.

Clayton W. Williams is as much a part of the Texas culture as cowboy boots and oil wells. This living legend has been an oilman, farmer, rancher, high-tech executive, financier, real estate developer, award-winning teacher, and gubernatorial candidate. His name is synonymous with success, adventure, and high-stakes business gambles that more often than not turned out successful.

Williams, chairman and CEO of Clayton Williams Energy, Inc., a public oil and gas company, was born in the small West Texas town of Alpine and grew up in Fort Stockton. He graduated from Texas A&M University in 1954 with a degree in animal husbandry and served two years in the U.S. Army. Williams used his abilities as a leader and can-do problem-solver to become one of the state's most successful entrepreneurs in the hallmark industries of Texas.

His successful business career began in 1957. With $2,000 saved from selling insurance and waiting tables in Mineral Wells, Williams and a friend, John May, founded May and Williams, an oil and gas company. May and Williams struggled until 1959 when the company drilled its first successful oil well. During the next four decades, Williams capitalized on that success, forming eight oil and gas companies in the state. His most successful company, Clajon, founded in 1961, became the largest individually owned gas company in Texas.

In the 1960s and '70s, Williams was one of the top natural gas wildcatters in the state, but it wasn't until the mid-1970s that he achieved the financial status for which he is known. That came on New Year's Eve 1975, when the Gataga No. 2 well blew out with such force that the nearby town of Mentone had to be evacuated. The Gataga, one of the largest wells in the Permian Basin, produced 32 million cubic feet per day.

In the 1980s and '90s, Williams became one of the top oil producers in the Austin Chalk Trend. In May 1993, Williams took his oil and gas company, Clayton Williams Energy, public on the NASDAQ stock exchange where it still trades today.

In recent years, he once again switched his focus to natural gas just as prices began to climb.

Williams received the Lifetime Achievement Energy Award in 1997 for a lifetime of bringing the good life to Texas and Texans through his many businesses. Other awards include the Distinguished Alumni Award from Texas A&M University, induction into the All-American Wildcatters, the Permian Basin Petroleum Association's Top Hand Award, 2002 Permian Basin International Oil Show Honor, an award from the Association of Building Contractors, and the All American Wildcatters Outstanding Wildcatter Award.

Among his many accomplishments, Williams is proudest of his lasting partnership with his wife, Modesta, and his five children and four grandchildren. He's a member of the First Presbyterian Church in Midland and supports the Midland community through civic and charitable organizations.

Gravity Map Service is the corporate descendent of Neese Exploration Company, formed in 1952 by Urban Earl Neese, who began his career with Stanolind Oil and Gas (now known as BP Amoco) as a scientist in torsion balance surveys, the forerunner to gravity meter surveys.

In 1948, Neese formed Tidelands Exploration Company of Houston and Calgary, followed by Neese Exploration Company to contract gravity surveys both domestically and internationally. Neese had field offices throughout the world, with headquarters in Houston.

Neese organized Gravity Map Service in the 1970s to acquire and provide a library of precision gravity data that would be obtained by Neese Exploration and be available for licensing by oil and gas companies. Gravity Map Service's digital data set includes proprietary gravity data throughout the Mid-Continent, Rocky Mountain, West Coast, and Gulf Coast regions. Gravity Map Service data integrated with other forms of geophysical and geological information creates a valuable tool for analyzing the subsurface in the search for hydrocarbons.

In the 1970s, Urban Neese's son, Robert, joined his father to expand gravity services. Robert Neese formed Neese Instrument Company to provide geophysical instrument rentals that include the Worden and LaCoste & Romberg Gravity Meters.

In 1990, Robert Neese formed Worden Gravity Meter Company to continue to manufacture and service the Worden Gravity Meter. Texas Instruments previously acquired the product from its inventor, Sam Worden, and operated it for over thirty-six years. Halliburton Geophysical Company bought the Worden product line from Texas Instruments. In 1990, Halliburton's Worden division was transferred to Robert Neese. Dennis Rohan, Charles Pustejovsky, Don Boone, and Johnny Gonzales manage the Worden group. Juan Hernandez was with the company until his death in 2001.

Urban Neese continued to be active in the gravity services group until his death in 1996. Robert Neese, the company's managing director, assumed the role of running the company several years prior to his father's death.

Like all successful companies, Neese relied on several key individuals throughout the years. Reed Kubena, chief geophysicist and office manager; Harold Burge, field operations manager; Darrow "Bud" Warren, surveyor; Luis Lanners, party chief; Matt Busha, surveyor and meter operator; Vertrice Jaffe, executive secretary; Wesley Pustejovsky, party chief; Dr. John Sumner, consultant; Bob Gardner, data processor; John Sorrell, gravity specialist; Betty Mumford, mapping; Maurice Kovar, data processor; and Mac McCord, surveyor, contributed greatly to the company's success.

Others who contributed to the company's growth are Kay Crockett, marketing; Ken Stupka, computer consultant; Dr. Joe Ku, gravity and magnetic consultant; Betty Hansen, mapping manager; William Krum, convention exhibit support; Richard P. Swirczynski, consulting geologist; and Charles "Preacher" Myers, seismic consultant.

Gravity Map Service and the Neese Companies are headquartered in Richmond, Texas, southwest of Houston. Its representatives are located around the world. Robert Neese continues to operate the gravity service group with a team of dedicated professionals utilizing the latest in earth science technology. The mission of the Neese group of companies continues to be to provide their clients with quality data, services and equipment through professionalism and integrity. They are members of the American Association of Petroleum Geologists and the Society of Exploration Geophysicists.

GRAVITY MAP SERVICE

NEESE & WORDEN

Left: Urban Earl Neese, founder of Neese Exploration Company, the predecessor of Gravity Map Service.

Below: Robert Neese.
COURTESY OF JOSEPH DYBALA PHOTOGRAPHY, INC.

Bottom: A Worden Gravity Meter being operated by Thomas Schultz.

James E. Smith

Above: James E. Smith.

Right: James E. Smith at work on location with clients.

James E. Smith is known worldwide for more than forty years of diversified petroleum engineering experience. After graduating from Texas A&M University in 1958 he began his career as a regulatory engineer with the Texas Railroad Commission, where he worked his way up through every engineering position the Oil and Gas Division offers.

Notable accomplishments included elimination of pollution practices associated with open pit and shallow well disposal, as well as planning, supervising and developing Texas' "Orphan Well" Plugging Program. The first state-funded plugging program in the United States, it was later adopted by other states. He was primary investigator and engineer during the East Texas Field slant well controversy. Smith personally planned and supervised re-entry and clean-out of over two hundred wells that had been intentionally junked to prevent being surveyed.

Smith developed the Railroad Commission Field Office operations organizational structure that has become a guideline for similar agencies. The landmark study of the Commission's goals and objectives that he developed became the basis for current field operations. He was principal investigator and developer of the Railroad Commission's "Lost Radioactive Source" regulation, which became a standard. Smith is best known as principal author and developer of the Railroad Commission Hydrogen Sulfide (H2S) Rule 36, which became the world-wide guideline for H2S operations.

After eighteen years as a highly respected regulator, Smith entered private industry. James E. Smith & Associates, Inc. (JES) rapidly became involved in well planning, supervision and general operations of over 1,000 wells of varying depths from 1,000 feet to 22,000 feet. In addition to involvement in drilling, completion and operations, he has developed a reputation as a reserves and evaluations engineer.

The author of many articles in industry publications, he has also held numerous leadership positions with professional engineering organizations. Hydrogen sulfide operations have long been a major interest. He recently emerged as a leader in the developing field of acid gas disposal by underground injection.

Smith has long demonstrated a special enthusiasm and interest in new developing technologies. He is known by the industry as the "ultimate generalist" for his wide knowledge of today's oil and gas engineering practices and his acquaintance with experts in varying fields. Smith is now regarded as a highly effective and experienced expert witness in oil and gas litigation issues.

He is active in civic and business pursuits. His family includes his wife, Mary, a retired elementary teacher; son, David, a professional engineer; daughter-in-law, Pangsri; daughter, Paula, a dedicated educator; her husband, John Temperilli, an environmental specialist; and several grandchildren.

Now in his sixties, Smith remains active in many phases of the oil and gas industry, continuing to develop and expand his technological interests. He recently commented that this is a great time to be an engineer in the oil and gas industry, adding that he wishes he had another forty years.

Parallel Petroleum Corporation is a publicly traded (Nasdaq: PLLL) independent energy company based in Midland, Texas, primarily engaged in the acquisition, development and production of oil and gas using enhanced oil recovery techniques including 3-D seismic technology in the Permian Basin of West Texas, East Texas, and onshore Gulf Coast of South Texas.

Founded in 1979, Parallel Petroleum has historically been an exploration company focusing on 3-D seismic, trying to make an oil or gas discovery that would increase its stockholders' value immediately. In the second quarter of 2002, Parallel sold its First Permian, L.P. assets, realizing a $31-million net gain, and changed its business plan to focus more on lower risk acquisition and development of producing properties and less on higher risk exploration. The company changed its business plan because of the risk associated with being solely an exploration company and its related boom and bust cycle. During the past ten years, exploratory reserve targets have become smaller, deeper, and more expensive, as confirmed by the industry's average finding cost last year of approximately $10 per domestic barrel of oil equivalent (6 MCF = 1 BOE).

Today, Parallel's primary focus is the acquisition and enhancement of producing oil and gas assets. To a lesser extent, the company will continue to drill lower risk exploratory oil and gas prospects associated with 3-D seismic. In December 2002, Parallel closed its $46-million Fullerton acquisition of long-life, shallow oil properties located in the Permian Basin of West Texas. The company expects these properties to have a lot of high-quality, inexpensive, upside potential with enhancement opportunities that have a higher probability of success than drilling exploratory wells. This acquisition was an extremely significant event for Parallel and will be a "cornerstone" asset providing substantial long-term cash flow and development opportunities.

Parallel's primary objective is to achieve strong growth in net asset value per share by capturing the unrecognized value associated with oil and gas assets through the efforts of its multidisciplined team. Parallel's staff, six outside board members, and outside consulting experts operate as a team that leverages its intellectual assets to recognize opportunities and execute its defined business plan. Parallel's business plan is to acquire producing properties and focus on projects that have economic impact, emphasizing exploitation, and enhancement activities in addition to exploration. It also seeks to expand the scope of its operations by diversifying its development and exploratory efforts, both inside and outside its current areas of operation, focusing on established geologic trends in which it can use the engineering, operations, financial, and technical expertise of its entire team. Parallel Petroleum is located on the Internet at www.parallel-petro.com.

Above: Corning Loudamy No. 1 Canyon Reef Well, Howard County, Texas was discovered in 1993.

Below: Layden-Yendrey No. 1 Yegua Gas Well, Jackson County, Texas was discovered in 1998.

Parallel Petroleum's common stock is traded on the Nasdaq National Market System (symbol: PLLL).

MIDLAND REPORTER-TELEGRAM

The Midland Reporter-Telegram *is located at 201 East Illinois in Midland, Texas.*

The *Midland Reporter-Telegram* is a Hearst Corporation newspaper located in the heart of the fifty-four-county Permian Basin. It has a daily circulation of more than 20,000 and a Sunday circulation of around 24,000.

The *Reporter-Telegram* came into existence in 1936 with the merger of the *Midland Daily Telegram* and the *Midland Reporter*, although they weren't the first newspapers to serve the Midland area.

The Midland Enterprise holds the honor as the area's first newspaper. It began publication in 1884 and closed eleven months later. Three others tried their hand at the news business before the *Livestock Reporter*, the predecessor to the *Midland Reporter*, began publishing a weekly paper in 1899.

It was that same year that Charles Caldwell Watson came to Midland. The printer and newspaperman arrived in Midland with no money after becoming the victim of a pick-pocket in Fort Worth.

He decided to pay a visit to the *Livestock Reporter* and found the publishers, Burt Rawlings and Virgil Albritton, getting ready to quit. Watson agreed to take over the paper on a three-month trial basis and soon discovered that the real publishers were a group of thirteen prominent cattlemen.

Eventually, the cattlemen who had formed the Midland Publishing Company turned over the *Livestock Reporter* to Watson, who renamed it the *Midland Livestock Reporter*.

Watson moved his printing plant several times in the early days and, according to a May 21, 1951, article in the *Midland Reporter-Telegram*, was "burned out twice."

Competition came and went in those days and several times Watson absorbed his competition. Then, in August 1924, a man named T. Paul Barron bought the *Midland Reporter* (it had dropped the word "Livestock" in 1903) from Watson and converted it to a semi weekly.

Watson moved to Belton and published the *Belton Journal*, but returned to Midland to publish the city's first daily newspaper. The *Midland Daily Telegram*, established in March 1927, operated under Watson's ownership until March 1929 when Barron, who's *Midland Reporter* had since become a daily, purchased it.

The *Daily Telegram* and the *Midland Reporter* merged in July 1936 to form the *Midland Reporter-Telegram*. James N. Allison purchased the newspaper in 1940 and the Allison family published it until 1979 when the Hearst Corporation bought it.

The newspaper converted to offset printing on November 11, 1975, and on April 2, 1990, switched from an afternoon newspaper to a morning newspaper as part of a nationwide trend.

Today, the *Reporter-Telegram*'s special coverage includes the Permian Basin Oil Report, a weekly section devoted to news of the oil and gas industry.

Shoreline Gas is an independent energy company established in 1990 to provide natural gas supply, gas gathering management and marketing services to producers, consumers and wholesale markets across the United States.

In this time of corporate mergers, acquisitions, and downsizing in the natural gas industry, the company headquartered in Corpus Christi has maintained its commitment to customer service and earned a reputation for quality, integrity, reliability and excellence.

Shoreline Gas assists producers, consumers and wholesale markets in the complex and volatile deregulated natural gas industry. Through experience, diverse business relationships, strategic alliances, and extensive resources, Shoreline Gas offers customers reliable gas purchases, transportation and supply.

The company's reputation for excellent service is best illustrated by the significant increase in the volume of gas marketed each year. Shoreline Gas fulfills the diverse needs of its customers by purchasing natural gas from independent producers, providing a variety of services including researching market options and evaluating pipeline alternatives as well as expediting new connections.

Shoreline Gas also offers services to the producer interested in outsourcing their entire gas management process. The company's team uses state-of-the-art technology for futures and weather information as well as electronic bulletin boards for gas transportation and scheduling.

Shoreline Gas purchases and markets natural gas throughout the United States on both interstate and intrastate pipelines accessing a diverse range of markets. Producers can take advantage of a wide range of services provided by Shoreline Gas. Those services include marketing, strategic analysis, contract administration, transportation management, accounting, and gas balancing.

Shoreline Gas offers consumers and wholesale markets the convenience of purchasing aggregated volumes from a single supplier. This ultimately maximizes the efficiency of any gas supply program and, by purchasing gas in a wide range of regions from leases operated by independent gas producers, markets are provided with both diversity and reliability.

The company offers swing, short-term, and long-term supplies on nearly all the interstate and intrastate pipelines located in our producing regions. Customers are also provided with a wide range of supply, management, and administrative services.

In addition, Shoreline Gas Field Services East Texas LP operates a natural gas gathering system located in East Texas. This system connects producers in the area with reliable markets as well as offering access to new markets. Expansions to the system as well as additional acquisitions are planned.

Shoreline Gas is committed to its role of producer advocate and its history shows the market presence necessary to perform has been successfully established. It is not without good reason Shoreline Gas has been ranked among the top gas marketing companies in North America in producer satisfaction surveys.

SHORELINE GAS

Above: (From left to right) David Musgrove, Donna Wright, Kyle Smith, Tammy Garcia, and Mark Miller.

Below: (From left to right) Chuck McGaughey, Leo Verrett, Tracy Skloss, and JoDean Murphy.

REICHMANN PETROLEUM CORPORATION

Richman Petroleum Corp. was incorporated in January 1994 under the direction of Dyke R. Ferrell and Erik Doughty, both from Corpus Christi, Texas. Erik had been in the oil and gas business since the 1970s and Dyke was a recent graduate of Southern Methodist University School of Business.

In April 1994 the company drilled its first well in Luling, Texas, and embarked on a series of four dry holes until its first field discovery in the fall of that same year. In 1994 the company was awarded the distinguished "Texas Discoverer Award" by the Railroad Commission of Texas for the discovery of the Hemlock Field in San Patricio County and the Kittie West Field in Live Oak County.

The company continued to drill exploration and development wells until 1998 when oil prices dropped to $10 per barrel. At that time, the company began buying production and accumulated more than one thousand barrels of oil per day equivalent by the time oil was back up to $30 per barrel the following year.

In 2000 the company began drilling deeper exploration wells in South Texas by embarking on an eighteen-thousand-foot wildcat. In August 2000 the company successfully drilled a record 17,800-foot well in Kenedy County, Texas on the Armstrong Ranch in 484 rotating hours.

Today, Richman Petroleum Corp. is still an independent oil and gas exploration company drilling deep wells in South Texas and pioneering new technology in the discovery of oil and gas reserves. The company celebrated its ten-year anniversary by changing the spelling to "Reichmann Petroleum Corporation" to honor the original German spelling of the last name of Ferrell's grandfather, Carl Russel Reichmann.

In 2001 the company began participating with Discovery Geo in developing over five hundred thousand acres in New Zealand. That same year, the founders of the company teamed up with Charlie Lawrence and began Texas Wyoming Drilling with one small rig in the Powder River Basin of Wyoming. By 2003, Texas Wyoming Drilling had nine rigs running in both states drilling coal bed methane gas wells in Northeast Wyoming and Barnett Shale wells in the North Texas area.

From the inception of Richman Petroleum in 1994 through the end of 2003, both companies had drilled more than 1,000 wells and continue to drill more than 75 wells per year.

Above: Dyke R. Ferrell.

Below: Richman Petroleum's corporate office is located at 223 College Street in Grapevine, Texas.

Texas Alliance of Energy Producers, with offices in Wichita Falls, Abilene, Houston, Austin, and Fort Worth, represents the interests of the oil and gas industry, ensuring that tomorrow's economic climate will allow its members to grow and prosper.

The Alliance was created in 2000 through the merger of the North Texas Oil and Gas Association and the West Central Texas Oil & Gas Association. Its combined membership of more than 2,000 comes from 187 cities in 21 states.

Some 85 percent of members are owners, co-owners or chief executive officers, and 79 percent are in the exploration/production business. Their common purpose is to protect the oil and gas industry by speaking with one voice at both the state and federal levels of government. Alliance members join together to help enact public policy designed to strengthen America's independent oil and gas industry.

The Alliance's presence before the Texas Legislature, Texas Railroad Commission, Congress, and federal regulatory agencies assures that the concerns of independent oil and gas producers are foremost in the minds of legislators and bureaucrats.

The Alliance's professional staff of lobbyists allows it to make the personal and frequent contacts needed to influence public policy and Alliance experts provide testimony showing the potential impact of proposals on the bottom line.

Its grass roots lobbying program can be mobilized to contact key legislators and member delegations of volunteers, led by the Alliance staff, meet with prominent lawmakers.

Coalition building and the Alliance's political action program that seeks to create a more favorable political and economic climate are prime examples of the power of unity that the Alliance uses to help its members. Members can also purchase a wide range of insurance policies and sales and property tax services and have access to a program that reduces electrical costs.

Members receive access to legislative and business information through newsletters, a monthly magazine, and insider forums and staff briefings. *Newsline* is a monthly newsletter that covers government relation's issues important to members. *Supply & Demand*, a weekly newsletter, focuses on supply and demand issues for crude oil and natural gas.

Exec Report, another weekly newsletter, focuses on management and employee issues of specific interest to industry executives, and *The American Oil & Gas Reporter* is a fast-paced monthly magazine that covers the industry from A to Z.

The Alliance seeks to focus public attention on the crisis facing the domestic oil and gas industry, fights negative images that make voters unsympathetic to problems facing independents, and takes a stand against radical environmentalists.

Through its staff and member volunteers, the Alliance plays a significant role in political battles in Austin and Washington and in the public relations battle in the court of public opinion.

TEXAS ALLIANCE OF ENERGY PRODUCERS

Above: Roy Pitcock, Jr. (left), chairman of the board, Texas Alliance of Energy Producers, and Charles W. Seely (right), immediate past chairman of the board, Texas Alliance of Energy Producers.

Below: Senator Kay Bailey Hutchison and Alex Mills, president, Texas Alliance of Energy Producers.

PHOTOS BY JAY B. RUSOVICH

Joe B. Foster, former chairman of Tenneco Oil Company, and twenty-six former Tenneco employees founded Newfield in 1988. The company was capitalized with $9 million by an investment group led by Charles Duncan, the University of Texas endowment fund, and personal investments from the founding employees. A second private placement in 1990 added $37 million in capital from investors including Yale and Duke Universities, Dartmouth College, and Warburg, Pincus Investors, L.P. In November 1993, Newfield completed its initial public offering of common stock and began trading on the NYSE under the symbol "NFX." The company went public at a split adjusted price of $8.75 per share.

Although Newfield had no properties at the time it was founded, the company had a solid list of business principles and an eye on creating value in a seven-hundred-block region of the Gulf of Mexico. Today, Newfield ranks as one of the largest producers in the Gulf of Mexico with more than 150 production platforms and nearly 250 lease blocks in both shallow and deep waters.

The original group of 26 employees has grown to about 500. Newfield's success in the Gulf of Mexico allowed the company to establish new focus areas onshore. Although about half of Newfield's reserves and production are still attributable to its Gulf of Mexico operations, the company has a significant footprint in South Texas and the Mid-Continent. The expansion into these regions began in the mid-1990s. Following several large acquisitions, Newfield now controls nearly four hundred thousand acres in South Texas and is running an active drilling effort. In the Mid-Continent, Newfield acquired longer-lived reserves in 2001 and is actively working to grow this asset base. Other new areas of activity for the company include the Gulf of Mexico's deepwater and the North Sea.

Newfield has enjoyed rapid growth and has created significant value for its shareholders. Today, the company's enterprise value is in excess of $3 billion.

Newfield is committed to protecting the environment. In all of the company's areas of operation, it strives to protect the health and safety of its employees and the communities in which it operates. Each of Newfield's employees, contract consultants and contractors has a responsibility to contribute to a safe and healthful workplace and protect the environment.

Rimkus Consulting Group, which is headquartered in Houston, utilizes a multidisciplined staff of over 300 professionals in 22 cities worldwide to provide business and technical consulting services to insurance companies, attorneys, corporations and other clients. Rimkus has roots in industrial construction, and its first assignments revolved around surety claims. When a contractor failed to meet his contractual obligations (due to bankruptcy, for example), Rimkus aided the surety bond provider by assessing the work completed, contractual obligations, and the cost of completion by another contractor. This was a common problem for energy industry businesses during the mid-1980s.

Rimkus' business then expanded to include all construction-related subjects and all aspects of the oil and gas industry—geology, drilling and production, pipeline construction and operation, marketing, and contractual issues. Building upon the construction experience, Rimkus expanded to address general property problems (such as roofing, foundations, water intrusion, and mold), industrial problems, explosions, industrial hygiene, and most recently, forensic accounting.

Robert E. Rimkus founded the company in 1983 and was joined in 1987 by Gary W. Markham, S. Frank Culberson, and Ralph S. Graham to form Rimkus Consulting Group, Inc.

All except the late Robert Rimkus continue as key players in the company's management team. With its business expansion to address a wide variety of industrial, commercial, and residential problems, Rimkus developed a reputation as the "go-to" team for these kinds of analyses. The company receives 40 to 50 new jobs daily involving mishaps or problems, starting work "when bad things happen."

The company began opening branches in 1989 with the Dallas site, and now has 6 offices in Texas, 19 in the U.S., and 22 worldwide, with over 400 employees and 2002 revenues of $42 million—$28 million in Texas, and $14 million in Houston alone. It has handled approximately 33,000 new assignments for a wide customer base during the past five years.

Some of the high-profile Houston-related incidents in which Rimkus has been involved are explosions at Phillips Chemical, Arco Chemical, the Brenham Salt Dome, and the Oklahoma City Federal Building; the One Allen Center fire; the construction crane collapse atop the Fred Hartman Bridge, residential and commercial damages from Tropical Storm Allison; and the *Exxon Valdez* crude oil spill.

Rimkus is an active part of the community, through contributions to numerous charities, as a member of the Greater Houston Partnership, and as a benefactor of the Houston Livestock Show and Rodeo.

PAUL DeCLEVA

A native of Budapest, Hungary, Paul DeCleva completed Piarist Prep School and then attended the Ludovica Academia—the equivalent to West Point Academy in the United States. Wounded while fighting on the Eastern Front during World War II, he served thereafter in the Hungarian Royal Guard. His last assignment in Hungary was as military aide to the Speaker of the House.

During the rearrangement of Central Europe by the victorious powers, Hungary was assigned to the Russians, who established a Communist regime with its associated terror. Under the new order, DeCleva was designated a "class enemy" of the system and was to be liquidated. Despite complications, he managed to leave the country and spent six months in Sweden before entering the United States in February of 1949.

After spending a year in New York, he decided to move to Midland, Texas, to reestablish himself in some fashion. He was hired as a draftsman with Atlantic Richfield, the predecessor of ARCO, which was his introduction to the oil business. When the 1952 steel strike shut down all the steel mills in the United States, oil operators in Midland were unable to obtain tubular goods and were forced to curtail exploration activities. Through his European connections, DeCleva managed to locate pipe and import respectable quantities, which were delivered to the operators in a timely fashion. This took him into the oil fields, allowing him to further familiarize himself with the oil business.

Remembering from his drafting days that the Westbrook Field in Mitchell County was not defined by dry holes, he leased a good portion of the east part of that field. Within two years, he had drilled thirty-two producers on this trend. He later extended his holdings to the north, where he drilled wells in the Glorietta and San Andres Fields in Scurry County. He also acquired leases in the Monahans area to prospect for Queens and Yates Sand production, which was very profitable, and engaged in secondary recovery in the same area. The Sealy Lease has a record of 8,000 barrels recovery per acre. Needless to say, this was significant even when oil prices stood at just $3.00 per barrel.

DeCleva's later activities were extended to the North Texas region, where he developed both primary and secondary recovery production. Currently, Paul DeCleva and related companies sell approximately thirty thousand gross barrels of oil per month.

DeCleva expresses pride in the fact that during these years he has been able to contribute in a small way to the national economy of hydrocarbons, so the American public has had availability to the cheapest oil in the whole world.

Paul DeCleva quotes Stalin in saying, "The death of one person is a tragedy, but the death of a million people is a statistic." DeCleva escaped being in the millions and says that even though he has endured the pain of a lot of dry holes, "I still feel that I made Stalin turn over in his grave."

Paul DeCleva

The *Odessa American* and what now is known as Freedom Communications Inc. have been partners since 1948 but the roots of Odessa's newspaper can be traced back to 1895.

Since August 13, 1948, the *American* has published millions of words about Odessa and the Permian Basin, chronicling the area's transition from pre-eminent supplier of oil and gas in the United States to the increasingly diversified retail, medical, cultural, and education center of today.

But thirty-two years before *Odessa* was incorporated as a city, the first official newspaper in the area, the *Odessa Weekly*, hit the newsstands. Started in 1895 by "Uncle" Billy Griffin, the publication survived just one year, but soon was followed by a string of other weekly publications.

In 1940, Henderson Shuffler combined two competing publications to form the *Odessa American*, which he later sold to Lone Star Newspapers, Inc., a publishing company. The paper then was sold again to Ridder Publications.

Since Freedom acquired the *OA*, only six publishers have led the newspaper. The first was the late V. L. DeBolt, who moved his family from New York in 1948 when he was named publisher. DeBolt died in 1974 and was succeeded by one of his sons, Lyle, who guided the *OA* through 1979. Lyle DeBolt was followed as publisher by David Lyons (1980-1990), Ray Stafford (1990-1994), Bill Salter (1994-2003) and current publisher Patrick Canty, who came to Odessa in June 2003 when Salter was named publisher emeritus.

Each publisher has found himself leading a constant parade of technology as almost unbelievable changes have swept the industry since 1948.

In 1952 the *OA* office was moved from Fourth Street at Texas Avenue to its present location at Fourth Street at Jackson Avenue. And the original press was replaced during the 1950s, which allowed for more newspapers to be printed in a shorter time period.

During the 1960s, the newspaper's staff doubled as its circulation increased and the building was expanded to house the necessary operations. But the biggest changes came as a result of the computer age. Since becoming part of the Freedom family, the *American* has gone from the old-style printing utilizing Linotype machines and masses of lead and individual lines of type to computerized processes that allow a page to be laid out and read to be put on the press in a matter of minutes.

Above: The Odessa American *employs the latest production technology.*

Below: The Odessa American *earned a Pulitzer Prize for photography in 1988 for Scott Shaw's photograph of Jessica McClure being lifted from a well in Midland.*

The biggest feather in the *OA*'s cap came when it won its first and only Pulitzer Prize (for photography in 1988). The image that won the coveted prize was a compelling photo of toddler Jessica McClure being lifted from a Midland well after a rescue operation that attracted worldwide attention. Scott Shaw, an *OA* staff photographer at that time, took the photo.

In the ensuing years, the paper has switched from an afternoon newspaper to a morning publication (as of January 1, 1990) and has made increasing use of color and digital photography. The paper also has developed a web presence by creating an on-line edition that has attracted many Internet users.

Considering the innovations that have swept the *Odessa American* since it was acquired by Freedom, its managers and staff can only imagine what the next half-century might hold.

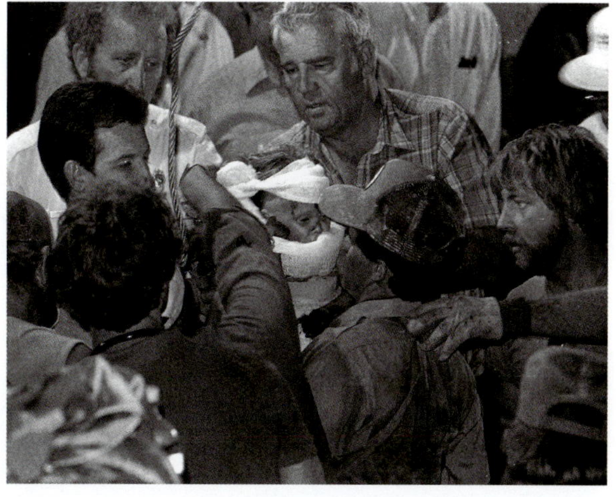

With a history dating back to 1973, two Texas Tech University geology graduates, Richard F. Spencer, class of 1957, and Osler C. Hutson, class of 1956, founded Sha-Jam Operating Corporation They originally combined their exploration expertise under the name of Spencer & Hutson, and later changed the company's name to Sha-Jam.

Hutson began his career with Humble Oil and Refinery Company where most of his efforts were concentrated in southeast New Mexico. Spencer, whose career began with Amoco, explored for hydrocarbon accumulations at various locations in Texas, New Mexico, and Oklahoma.

Sha-Jam concentrates on the exploration and production of hydrocarbons in Texas and New Mexico, including the generation of oil and gas prospects, securing of leases, acquiring funds for drilling, and completion, producing, and selling the product.

During the company's early days, Sha-Jam discovered significant oil and gas fields in southeast New Mexico, including the Humble City Strawn Field, located near the town of Lovington, and the Townsend Field.

Sha-Jam has generated many oil and gas prospects that have resulted in significant petroleum reserves impacting the city, state and country.

In the late 1970s and throughout the '80s, Spencer and Hutson initiated efforts to drill and expand the now gigantic Austin Chalk Field of Lee, Fayette, and Gonzales Counties in Texas. They drilled and completed many high-capacity wells. One of the largest of these, their No. 1 Hawn in Lee County, Texas, with potential open flow for 100 million cubic feet of gas per day, plus 10,000 barrels of oil per day.

Sha-Jam's five-person organization operates locations in Midland, LaGrange and Ballinger, Texas, with headquarters at 507 West Tennessee in Midland. The company continues to generate oil and gas prospects with good economic returns, primarily in the States of Texas and New Mexico.

Hutson and Spencer were drawn to one another as business associates by their common Christian faith and commitment to the cause of Christ. During the course of their more than thirty-year professional alliance, both of them have yielded time and resources to teach others ways in which to live the victorious life in Christ Jesus. Both men remain extremely active in church and related endeavors.

RIDGMAR OIL AND GAS COMPANY

The headquarters of Ridgmar Oil and Gas Company at 2601 Ridgmar Plaza in Fort Worth, Texas.

The history of Ridgmar Oil and Gas is really a family history, beginning with Sam Weiner, a Russian immigrant who was in Arkansas during the oil boom and followed it to West Texas. A tough, hard-drinking gambler who sired six children, Weiner went broke and set up a machine shop in Wink, Texas. Sam's oldest child, Ted, while a teenager, built a rig in the shop's backyard and began trading its use for pieces of oil deals. The first major success was named the "Weiner Pool."

When Ted's baby brother died of scarlet fever due to inadequate medical care, he moved the family to Fort Worth. He married Lucile Clements and they moved to Fort Worth for the birth of their first and only child, Gwendolyn. Ted opened an office in the old Fort Worth National Bank Building and founded the Texas Crude Company.

After the Korean War, Ted's brothers, Stanley and Charles, joined the business, with Stanley based in Midland operating the drilling company and Charles, a geologist, based first in New Orleans and later in Houston. In the 1940s, after the discovery by Texas Crude of the Spraberry Field, Ted built his office building at 2601 Ridgmar Plaza, the present headquarters of Ridgmar Oil and Gas.

Ted and Gwendolyn were always very close and he took her to meetings and on business trips, so she grew up with constant exposure to the business. She also had a great love for and fascination with the business.

Stanley's health began to fail in the early 1960s. The drilling companies were merged with the Fluor Corporation, giving all three brothers seats on the Fluor board. Stanley died in 1969, and in 1971, Ted and Charles went separate ways dividing the properties between the three families.

Ted started a new company, but his health began to fail in 1975 and he died in 1979, when Ridgmar Oil and Gas was two years old. Gwendolyn became president and owner. Ridgmar has since drilled in sixteen states and Canada.

The company has been blessed with excellent employees. Katheryn Votaw began working as Ted's secretary in 1950 and remained thirty-five years, eventually handling nearly every phase of the business. After Ted's death, she coordinated the estate settlement, transitioning with Gwendolyn into Ridgmar Oil and Gas. She retired in 1984 to be followed by her assistant Nina Pruitt who was with the company thirteen years until her retirement. Michael E. Bertrand, comptroller and general office manager, worked for the company for nearly fourteen years until his untimely death in 2002. Norma Reid and Veta Weatherby were also valued employees.

Ted was inducted posthumously into the Petroleum Hall of Fame in the 1980s with Gwendolyn accepting this honor on his behalf.

SIVALLS, INC.

Above: Shop building and employees in the Odessa service department finishing out a dehydration unit in 1965.

Below: C. Richard Sivalls inside the Odessa shop with Russian treaters in the background, Fall 2002.

With over a century of expertise and tradition, Sivalls, Inc., manufactures equipment designed to separate well stream components, making it a major contributor to the world's energy supply. The conglomerate stream of every oil and gas well includes water, sand, rock particles, hydrogen sulfide, carbon dioxide, and a long string of other impurities that must be removed before the oil or gas has any value.

The expertise behind Sivalls, Inc., began early in the twentieth century with James A. Sivalls, a "cooper" or barrel and tank maker for the oil industry. By the mid-1920s, his son, Charles T. Sivalls, was a partner in Black, Sivalls & Bryson, which produced steel-bolted tanks. When the partners sold out, Charles founded Sivalls Tanks, Inc. in Odessa, Texas, in 1947. The name was changed in 1977 to Sivalls, Inc.

The company's prime product was welded steel storage tanks for the oil industry in the West Texas-Southeastern New Mexico Permian Basin. In the late 1950s the Sivalls' third gene-ration, C. Richard Sivalls, a University of Oklahoma mechanical engineering graduate, began working to expand the Sivalls' product line and geographical scope. The company now has plants, offices, and representatives in principal U.S. and international oil and gas centers.

Control Equipment, Inc., a 1950 spin-off, is now a sister corporation with branches in Wichita Falls and Pampa, Texas. CEI distributes fluid flow parts and accessories to manufacture and maintain surface production equipment and a wide range of PVC and other plastic products.

A fourth-generation family member, Stephanie Sivalls Latimer, began working for the company in 1979, and is now vice president of administration.

Sivalls' central Odessa facilities were enlarged in 1979. In 1981 a new twenty-five-thousand-square-foot fabrication facility was built at Pampa, and in 1982 another manu-facturing facility in Brownwood, Texas. In 1984, Sivalls acquired a Rocky Mountain-based company, Oil & Gas Processors.

Another spin-off, Tectrol, Inc., based in Odessa, focuses on the design and production of chokes and level controls, primarily used as stock standards on Sivalls equipment. Tectrol later designed and developed a gate valve line to compete in the market primarily devoted to gas pipelines in the Northeast.

Utilizing the latest in technological advances, Sivalls' engineers, specialists, production, and sales representatives all continue to make important contributions to the oil and gas industry, as well as playing a vital role in equipment used by the cement, sulfur, ammonia, fertilizer, water treating, power generating, and other industries.

For over eighty-five years, the Robinson family has been associated with the drilling industry. T. L. Robinson drilled test holes for lead and zinc mines in northeastern Oklahoma before the beginning of World War I. During the Depression years, his son, Glenn O. Robinson, drilled municipal water wells for Oklahoma City. Later, Glenn and his two brothers, Clyde and Gale, were early wildcatters in Illinois and in areas around Seminole and Muskogee, Oklahoma.

The Robinson men came to West Texas in 1937 with their three spudders and began drilling operations in Nolan County. After several disappointing dusters, they moved the rigs to the Sharon Ridge area in Scurry County, and in 1938, the discovery well, the R. O. McClure No. 1, was successfully completed.

In 1947, Glenn's son, G. R. (Bob) Robinson came to Texas following his service in the European theatre of operations during World War II, and worked for a year as an engineer for Stanolind Oil and Gas in Oklahoma City. Glenn and Bob soon formed Robinson Drilling Company, which was headquartered in Colorado City, Texas. Glenn worked as landman, Bob served as the tool-pusher, and Myra, Bob's wife, was the bookkeeper and payroll clerk.

At first the new company used the spudders that were moved from Oklahoma, but these were later converted to combination rigs. In 1948 the first new rotary rig was added. The father and son team operated Robinson Drilling Company for twenty years, developing leases in Scurry, Mitchell, and Garza Counties, as well as operating five rotary rigs as drilling contractors. Glenn Robinson died in 1962 in Miami, Oklahoma.

In 1966, Bob Robinson relocated the headquarters of Robinson Drilling Company to Big Spring. Following his death in 1972 as a result of an automobile accident, Myra reorganized the company as Robinson Drilling of Texas.

The company has benefited from the experience and loyalty of many long-term employees. Chester G. Miller served as general manager from 1972 to 1982, and H. L. (Papy) Warneke, who worked for the company for forty-five years, was general manager from 1982 to 2002. Ray Alexander, office manager and comptroller, has been with the company for more than thirty years.

R. M. (Mike) Robinson, the son of Bob and Myra, assumed the position of general manager in 2002.

Through four generations of the Robinson family and over five decades, Robinson Drilling has drilled over fifteen million feet. The company operates rigs capable of depths from 7,500 to 13,000 feet and primarily operates within a hundred-mile radius of Big Spring.

ROBINSON DRILLING OF TEXAS, LTD.

A modern drilling rig near an old retired cable tool rig.

Texas Energy Museum

The Texas Energy Museum opened in January 1990 but its roots go back many years before. The Museum is the combination of two earlier museums—the Western Company of North America Museum in Fort Worth and the Spindletop Museum of Lamar University in Beaumont. The Texas Energy Museum is located at 600 Main Street in Beaumont.

The brainchild of H. E. (Eddie) Chiles, the Western Company Museum was created in 1979 to illustrate petroleum geology and the historical and current technology of petroleum engineering. The Western Company, which he founded in 1939 with two trucks and three employees, grew into a major oil equipment and service company and offshore drilling rig producer with revenues exceeding $700 million annually. However the company collapsed into bankruptcy in 1988, and Chiles began dismantling his operations. He also sold the Texas Rangers baseball club in 1989, but not before signing Nolan Ryan. At the same time, he donated the company museum to an interested coalition of Beaumont city officials, Lamar University regents, and private citizens.

The City of Beaumont offered to construct a facility, raise an endowment for operating funds, and combine the Chiles' collection with that of the Spindletop Museum. The Spindletop Museum opened in 1951 inside an exposition hall at the current site of the South Texas State Fair in commemoration of the fiftieth anniversary of the discovery of oil at Spindletop.

The exhibition added artifact donations to those of the Lucus Gusher Monument Association that had been created in the 1940s. In 1971 the aggregation found a new home in the former general offices of the Sun Oil Company. Lamar University assumed control of the museum in 1975, and with state funding opened a new Spindletop Museum in one wing of a former elementary school building on its campus in 1976.

At the same time, local citizens were working with the university to build an outdoor re-creation of the 1901 oil boomtown, Gladys City. The Gladys City Boomtown complex exists today as a sister institution to the Texas Energy Museum. Funds raised by the university during the 1970s to construct a modern petroleum museum adjacent to Gladys City were presented to the Texas Energy Museum project.

With support from local individuals, local and state foundations, petroleum companies such as Texaco and Chevron, Lamar University, and the City of Beaumont, the Texas Energy Museum was created in 1987.

Today, the Texas Energy Museum tells the story of oil through colorful, state-of-the-art exhibits. The exhibits depict petroleum geology, formation of oil and gas, history and technology of oil production, and refining and petrochemicals. Talking robotic characters relate their stories of early Texas oil drilling amid historical surroundings and working artifacts.

Guided tours by trained volunteers present the science of petroleum energy to school groups. Public educational programs for children and adults include Dinosaur Day, Family Discovery Lectures, Bubble Day, and Summer Science for Kids.

A casual observation made in 1904 by a geologist working for Theodore Roosevelt's United States Reclamation Service led to development of the largest natural gas fields in the world. Dr. Charles Gould, embarking on an extensive geological survey of the Texas Panhandle and New Mexico at the turn of the twentieth century, noticed large slabs of brown dolomite granite dipping dramatically into the earth along the Canadian River, north of present-day Amarillo.

Gould's task was to determine the water resources in the area. His observations led him to believe that dome-like structures might exist underground, natural traps for crude oil and natural gas. His study gathered dust on university shelves and in Washington archives for almost a decade before two Amarillo businessmen paid him a visit.

They were primarily interested in the possibility of finding oil on property they had leased in Oklahoma. Gould was skeptical. On the way out the door, one of his visitors asked if Gould happened to know anywhere in the Texas Panhandle that crude oil might be found. Almost as an afterthought, Gould said he thought the Canadian River breaks north of Amarillo might be a pretty good place to look.

In 1918, that became an understatement. Almost overnight, the northern Texas Panhandle was changed from depending on wheat and cattle to drawing tremendous resources from beneath the rocky soil. One gas well, discovered on the Masterson Ranch twenty miles north of Amarillo began to produce thousands of cubic feet of gas daily, enough to supply all the needs of the population well into the mid-1950s. Several 1918 vintage wells on the Masterson Ranch continue to produce, their flows unabated.

Those discoveries opened the floodgates to dramatic development of the Texas Panhandle gas field. By the mid-1920s crude oil was very much part of the picture, with large wells brought in almost daily. Many of the exploration companies relied on the research of Gould and his fellow geologists to increase the odds of success. This was the first time in the history of the U.S.

that scientific principles were used to mount a systematic search for hydrocarbons.

By 1922 more Panhandle men were working for the oil and gas operations than the cattle business. Within a few years the oil boom caused most towns north and east of Amarillo to triple then quadruple in population. In 1922 the City of Borger came into being within a two week period because of major crude oil finds along the Canadian River. Borger had a rough and tumble population, more concerned with cashing in on the black gold rush than in observing the law. It was the first town in Texas to be under martial law because of widespread lawlessness, verging on anarchy.

By the mid-'30s, Panhandle oil and gas production was so intense that complicated legal battles erupted between crude oil producers, natural gas producers, and those involved in "stripping" gas from crude oil flows. The "strippers" wanted the right to separate the high-Btu gas from the crude oil, process it though a gasoline plant and vent the dried gas to the air. Sour gas producers sought to force pipelines to buy their production.

A round of court battles and administrative hearings before the Texas Railroad Commission in the 1930s established the principal of correlative rights, common purchaser and market demand. Those issues are still much in the minds of producers, gatherers and pipelines in the twenty-first century.

During the pre-war decade, hundreds of miles of gas gathering lines and transmission lines were created in the Panhandle, linking it with fuel hungry metropolitan cities in the Midwest. Today, millions of homeowners from Amarillo to Chicago make their morning coffee with fuel transported from Texas Panhandle fields.

From an international perspective, the abundance of helium in the Texas Panhandle gas supply was the most significant development. The high concentration of the valuable gas made Amarillo the center of helium development and marketing on the globe. Even today, the Cliffside Field north of Amarillo produces the world's largest supply of helium.

PANHANDLE PRODUCERS & ROYALTY OWNERS ASSOCIATION

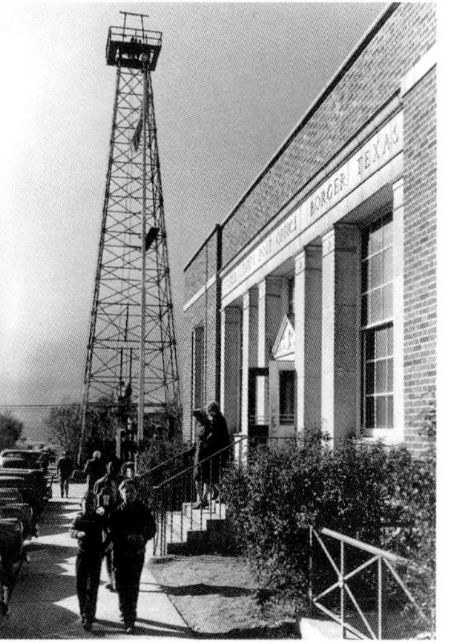

Borger's skyline has always been dominated by the oil and gas business. This steel derrick marked a drilling location next to the downtown post office in the early '40s.
COURTESY OF THE PANHANDLE PLAINS HISTORICAL MUSEUM.

CHIEF OIL & GAS

Chief Oil & Gas LLC is a relative newcomer to the industry. Founded in 1994 by Trevor D. Rees-Jones, Chief began slowly before finding its niche and developing a drilling strategy all its own.

Rees-Jones, the company's founder and president, worked for more than ten years as an independent oil and gas producer, drilling dry holes all over Texas before developing a strategy that proved successful.

Looking to lower his risk, Rees-Jones decided to refocus his efforts on the Wise County area of the Fort Worth Basin of North Texas, drilling infill development wells for the Bend conglomerate formation.

The company drilled thirty-nine Bend conglomerate wells over the next four years. Natural gas had been selling for an average of $2.00/MCF for the previous fifteen years, and the company believed it had a good future.

They were right, of course, albeit a little early in their forecast. With the stock market delivering unprecedented returns and with gas prices diving occasionally to $1.50/MCF, Chief found it difficult to raise money from investors.

But Rees-Jones and landman William G. (Bill) Kiker, his only full-time employee at the time, enjoyed working the techniques they had developed, figuring out, for instance, when another operator would let a lease go so they could sneak out and lease the acreage. Still, it was easy to get discouraged.

Drilling the Bend conglomerates was like picking the last bit of meat off the bone, and with gas at $2/MCF there wasn't much room for error. Besides that, the Fort Worth Basin was thought to be beyond dead—until the Barnett Shale play came along and got going in 1999-2000.

The company eventually shifted its focus from Bend conglomerates to the Barnett Shale, playing a hunch that a second prong of Barnett production could be drilled out in this area, and they began leasing in northern Tarrant County.

The company now holds more than 20,000 acres in the Newark, East (Barnett Shale) Field, with more than 10,000 acres in Tarrant County. The second prong of the field did exist and Chief took aggressive steps to prove it up.

Chief began drilling Barnett Shale wells in earnest in May 2000 at a pace of a well or two a month. It has now drilled more than 120 Barnett wells and is drilling at the pace of six to eight a month.

In 2000, after completing more than 30 Bend conglomerate wells as productive, Chief was producing 3,000 MCF/day. Today, Chief is producing some 60,000 MCF/day from more than 100 wells and is ranked among the "Fifty Most Active Operators" in the United States in terms of wells drilled on an annual basis.

Chief is now working four drilling rigs full time and drilling at the pace of more than eighty Barnett Shale wells a year with an annual drilling budget of more than $60 million a year. And the company that began with two employees now employs over 30 full-time and contract personnel.

And Chief Oil & Gas' plans for the future? Simple. Keep drilling.

Below: The officers of Chief Oil & Gas, LLC (from right to left): Trevor Rees-Jones, Bill Kiker, and Cliff Thomson.

Bottom: Chief Oil & Gas working two drilling rigs on its Bonds Ranch lease in Tarrant County.

The Texas Oil & Gas Association is a petroleum trade association that unites the state's oil and gas industry to work for a common good.

With roots that go back as far as 1917, the Association is the oldest and largest organization in the state to represent petroleum interests and continues to serve as the only organization in the state to embrace all segments of the industry.

The Association's members total approximately 2,000, some of whom are executives of 50 of the state's largest energy companies. Members represent every segment of the industry encompassing 29 states and 128 cities across the nation.

The Association's activities focus in three areas: legislation, regulation, and public/industry affairs. Association work is concentrated in these areas in order to be responsive to the ever-growing demands made on the industry by federal, state, and local governments and by the general public.

In order to do this, the Association turns to its strong member-supported committee structure to mobilize industry representatives with expertise in various fields to deal with specific problems.

The Texas Oil & Gas Association, with headquarters in Austin, is one of four affiliated divisions of the U.S. Oil & Gas Association. The others are the Oklahoma Mid-Continent Oil & Gas Association, with headquarters in Tulsa; the Louisiana Mid-Continent Oil & Gas Association, with headquarters at Baton Rouge; and the U.S. Oil & Gas Association, Alabama-Mississippi Division, with headquarters in Jackson.

The concept of an all-inclusive petroleum organization was realized on October 13, 1917, when the Mid-Continent Oil & Gas Association was formed in Tulsa, Oklahoma. One of the main reasons for the group's formation was to provide essential supplies of petroleum and petroleum products to the U.S. armed forces engaged in World War I to, as one member put it, "float to victory upon a wave of oil."

Membership in the Association is open to anyone with a vested interest in continued prosperity of the oil and gas industry in Texas. Members benefit from Association membership by lending their voices to the collective expertise that creates policy and controls their ability to make money.

Members also benefit from networking with industry leaders; remaining up to date on state and federal developments; realizing substantial savings through participation in training, continuing education, and safety instruction; and enjoying additional savings on long-distance phone bills and insurance premiums.

The Association produces *Texas Oil & Gas Today*, the association's quarterly magazine; *Petroleum Industry Digest*, a faxed monthly newsletter; *Legislative Digest*, published biweekly during the legislative session; and other advisory bulletins.

Members also remain current through the Association's information center and timely speakers at Association seminars and the Texas Oil & Gas Industry Issues Conference.

TEXAS OIL & GAS ASSOCIATION

WALTER OIL & GAS CORPORATION

Walter Oil & Gas Corporation ranks among the top independent oil and gas exploration companies operating in the energy-rich waters of the Gulf of Mexico. Thanks to its financial strength, technical expertise, and pioneer spirit, this Houston based enterprise is one of the most prolific natural gas producers in the Gulf and it consistently places among the top twenty companies in wells drilled there annually.

WOGC's history dates back to 1981, when Houston Oil & Minerals Corporation, which was founded by J. C. Walter, Jr., merged with Tenneco. Walter's energy track record enabled him to raise money from pension funds and other investors that he needed to launch a new enterprise: Walter Oil & Gas Corporation.

Initially, WOGC focused its efforts onshore along the Texas-Louisiana coast. Then in 1984 in a bold move for such a small company, WOGC became one of the first independents to begin operating in the Gulf.

"The move to offshore operations enabled us to build strong working relationships with the majors before our competitors could," says J. C. "Rusty" Walter III, who joined his father's company in 1982 and has served as its president and chief executive officer since 1989. WOGC's primary focus is prospect generation and exploratory drilling, with a mixture of low, medium and high-risk projects.

WOGC formed gas and oil trading companies in 1988 and 1990, respectively, to market its production and also handle production for third parties. In the volatile oil and gas pricing period over the last fifteen years, these companies have been valuable in achieving better than market average prices for WOGC's and its partners' production.

In 1990, WOGC became one of the first independents to explore internationally. "We employed the same strategy internationally that worked for us in the Gulf," Rusty Walter recalls. "We sought opportunities that were too small for the majors to pursue profitably." Most recently, the company has begun looking at exploration opportunities in the UK sector of the North Sea.

Walter attributes the company's success in part to its entrepreneurial spirit. "We are risk takers. We're willing to try new things," he said. Walter credits WOGC's team of highly qualified geologists, geophysicists and engineers, as well as the use of advanced technology, such as three-dimensional seismic and processing technics and the development and use of subsea completion and production technology, as important factors in the company's success.

"Since we are not a public company, we can make decisions based on economics rather than on public market conditions," he adds. "We can react quickly when opportunities arise."

WOGC prides itself on its ability to find oil and gas consistently at a lower-than-average cost per barrel equivalent. As Walter explains, "We've been profitable at times when others were struggling."

Over the past twenty years, Walter says, "WOGC has built a firm foundation for further expansion and growth and remains well-positioned to take advantage of new opportunities as they arise."

Vanco Energy Company is a privately held oil and gas company whose mission is to explore and develop oil and gas in deep-water worldwide.

Vanco had its beginning in Wichita Falls, Texas in 1952 when founder and President Gene Van Dyke became an independent producer and operator drilling wildcats in North Central Texas, the Texas Gulf Coast, and later in South Louisiana, where the company discovered more than two TCF of gas in the West Lake Arthur Field in 1968.

In the late 1960s, Van Dyke shifted his focus to international exploration. In 1973 the company acquired licenses in offshore The Netherlands, beginning a twenty-three-year history of success in the Dutch sector of the North Sea, which culminated in the discovery of the Rijn Field (P/15, 1982) and the Horizon Field (P/9, 1983), among others.

During the mid-1990s, Van Dyke again refocused his company. Africa was selected, due to a recognition of the vast exploration potential in Africa's deepwater water basins and a firm belief that the deepwater production technology being developed in the Gulf of Mexico would be applicable in the mild African climate. Vanco first acquired the Anton Marin and the Astrid Marin Permits, offshore Gabon in 1997. In 2001 the Vanco Gabon Group drilled four wells, setting a world record water depth with the Judy 1 well, drilled in 2,791 meters water depth.

This first deal led to the acquisition and evaluation of other licenses in offshore Africa, including Morocco (1998), Côte d'Ivoire (1999), Equatorial Guinea and Namibia (2000), Madagascar (2001), and Ghana (2002). Vanco has become a leader in utilizing state-of-the-art geophysical techniques in exploring frontier basins with water depths greater than 1,000 meters; the basins are often salt-related and always contain huge structures with giant field potential.

In 2004, several years of exploration work will culminate with the drilling of four deepwater wells in Africa, two of which will be operated by Vanco. High-technology geophysical work continues in Madagascar as Vanco operates the first three-dimensional survey ever conducted in East Africa.

Gene Van Dyke's determination to build Vanco into a major company is coupled with a dedication to good corporate citizenship. Vanco contributes to the communities where it does business by offering scholarships, professional training, and assistance with local social programs. Vanco also respects human rights, cultures and customs of its host countries, while working to ensure that operations are conducted in a safe and environmentally responsible manner.

Gene Van Dyke, founder of Vanco Energy Company.

Kendrick Oil and Gas Company was founded in 1918, during the Burkburnett oil boom by P.S. Kendrick, Sr., who was born February 3, 1890, in Kendrick, Mississippi. His mother died when he was three years old and his father when he was twelve. He lived with family friends and relatives until age seventeen, when he went to Albany, Texas, to become a cowboy.

Texaco brought in a gas well in nearby Moran in 1909 and a two-hundred-foot-deep discovery followed. In 1913, while earning $35 a month working for Albany National Bank, Kendrick bought his first oil and gas lease south of Albany, paying twenty-five cents per acre. He drilled a dry hole.

He quit the bank in 1918 and became an Internal Revenue Service agent stationed in Wichita Falls, south of Burkburnett. On weekends he scouted the area in his new Hupmobile, for which he had traded an oil and gas lease, hoping to get in on the Burkburnett boom. He called this "the roughest riding car ever built, but good to drive country roads looking for an oil lease."

There were no Blue Sky laws and the scheme was to sell stock in your venture, print certificates, set up a box on the sidewalk and sell stock. Kendrick bought shares in the Burk Waggoner Oil Company and a lease on ten acres from Clois Green on Bob Waggoner's

160-acre farm. Running out of money several times while drilling the well with cable tools, they had to stop drilling to sell more stock.

The well came in flowing 3,000 barrels a day at 1,700 feet, a discovery for the three-mile northwest extension of the Burkburnett pool. Kendrick drilled three wells on his ten acres. Oil was selling for $3.06 a barrel and gasoline for twenty-seven cents a gallon with no tax.

Kendrick sold out, bought a two-thousand-acre ranch southeast of Albany and started breeding Polled Hereford cattle. His bulls and steers won blue ribbons at fat stock shows in Fort Worth, Kansas City and Chicago. He served as president of the American Polled Hereford Association.

The Great Depression of the 1930s put him out of the cattle business and he continued promoting oil and gas leases and working as a cable tool drilling contractor owning three cable tool rigs, which he reluctantly traded for rotary rigs in the 1950s.

Kendrick Oil and Gas has been a family-owned company since 1918. In 1967, Phil Kendrick, Jr., left the family business to work for the investment banking firm of White, Weld and Co. in New York and Los Angeles. In 1973, Phil left White, Weld and Co. to start Harken Energy, taking the company public in 1979 and selling his position in 1983. He then became president of Kendrick Oil and remains in that position today.

Above: P. S. Kendrick, Sr.

Below: Phil Kendrick, Jr.

The Youngblood family traces its immediate history to J. Lee Youngblood, who was born in Columbus, Georgia, on January 12, 1909, the second child of Moses Snow Youngblood and Annie Alley Youngblood.

Not long after that, the family moved to Wewoka, Oklahoma, where Snow Youngblood owned a hardware store. After Snow Youngblood died, J. Lee Youngblood and his older brother, Laurence Snow Youngblood, built the Youngblood Hotel in Enid, Oklahoma, in 1928-29.

Shortly after, the two brothers founded the Youngblood & Youngblood Oil Company, a partnership, which was the first oil and gas brokerage firm. They became the largest brokers in North America, buying leases for various companies throughout the United States, Canada, and Alaska.

They gradually changed from brokers to operators, drilling wells either individually or in partner-ships with others. J. Lee operated an office that moved from one location to another in order to participate in the most active plays. He had an office in Illinois for several years, then moved it to Shreveport, Louisiana, and, finally, to Dallas in 1952.

When L. S. Youngblood died in June 1965, J. Lee established the J. Lee Youngblood Oil Company and operated the company from his Dallas, Texas, office. He continued to explore for oil and gas as an operator until the early 1980s, when he ceased operations and began participating in prospects operated by others. That continued until his death on January 29, 1996.

During his lifetime, J. Lee belonged to the Texas Independent Petroleum Royalty Owners (TIPRO) and served on various committees and boards, including Sabine Royalty. He also belonged to the Dallas Petroleum Club and served as its president in 1963.

He belonged to the Brookhollow Country Club; the Dallas Country Club; the Garden of the Gods Club in Colorado Springs, Colorado; and the Birnam Wood Club in Santa Barbara, California. His wife, Juanita Youngblood, was a tremendous help and support to him throughout his career. Their daughter, Penny Lee Youngblood of Dallas, now runs the oil and gas business as well as the Circle Y Ranch in Aubrey, Texas.

J. Lee Youngblood.

UNISON DRILLING INCORPORATED

Above: Annie Mello, owner of Unison Drilling Incorporated.

Below: Maurice Strickler's casket is loaded onto his Peterbilt truck to haul him to the cemetery in Moore, Texas.

Maurice Strickler and Annie Mello in Devine, Texas, established Unison Drilling, Inc., in March 1985. It is headquartered in Devine, where Annie was born and reared. As children, she and Maurice worked for their father's water well drilling business there and Annie received her "schooling" on drilling equipment and the language and terminology used around drilling rigs. The family's eight children—five boys and three girls—were all roughnecks on their father's drilling rig at one time or other.

When Annie and Maurice became partners in the drilling business they began searching for their first rig, driving to Oklahoma, Kansas, Arkansas, Louisiana, and Mississippi before finally leasing a rig from a Mississippi contractor. Nine months later they purchased their first rig from a Texas drilling contractor.

Maurice loved attending oil field auctions to buy drilling equipment. Annie sometimes accompanied him, though she didn't much care for the auctions.

"I just enjoyed being with him," she remembers. "Even though I was his sister, he would take the time to explain how something worked, or why it would not work. Once he sent me to an auction in Odessa by myself to buy a rotary. I thought I had made a good buy until we got it back to the yard and Maurice looked it over. He told me the only thing he would do with that rotary table was to set it in his front yard and plant flowers in it. After that, he never sent me to another auction."

When Maurice was killed in an auto accident in 1994, Annie arranged to have his body hauled to the little country cemetery in Moore, Texas, on the bed of his 1991 Peterbilt haul truck.

"I think we made the funeral director a little uneasy, but I knew Maurice would have fired me if I had hired a hearse to transport his body when we had plenty of trucks to do the job," she says.

After Maurice's death, Annie decided to carry on the drilling business instead of selling the two rigs they owned at that time. With the help of her husband, Anthony; her son, Michael; her brother, Dwaine; and her tool-pusher, Clement Wofford, Unison Drilling has continued to grow. The company now has three rigs drilling in the South Texas area.

"I don't plan on becoming a big drilling company," Annie says. "I want to stay small. I like knowing exactly where my rigs are working and knowing my employees by name and face. I like knowing every piece of equipment on my rigs and where every dollar is spent. "I can't wait to get to heaven so I can talk with Maurice. I think by now I could teach him a few things about the drilling business."

The Texas Independent Producers & Royalty Owners Association, the largest association of its kind in the nation, is committed to promoting the interests and welfare of independent oil and gas operators.

TIPRO works with interest owners, royalty owners, and businesses that provide services for the energy industry. It does this by providing outstanding legislative and public representation, access to valuable information and excellent opportunities to develop business relationships.

The group's origins lie in the struggle to develop the East Texas Field. In the 1930s, the independent producer had no effective voice of his own in the global realities of the petroleum industry. It became clear that there was an urgent need for creation of a state institution organized to address national as well as state issues. So, in March 1946, concerned parties formed TIPRO to create a formal structure of independents.

An excerpt from the December 15, 1947, *Independent News*, the official publication of the Texas Independent Produces and Royalty Owners Association, quoted from TIPRO's bylaws: "The purpose of this organization is to promote the interest and general welfare of the Texas independent producer and royalty owner of oil, gas, and other mineral rights."

Traditionally a lobbying group, TIPRO has in the past decade become increasingly active in technology transfer efforts among independent producers. It has helped educate independents in the fields of the oil patch as well as the halls of government, both at the state level and, after World War II, in the U.S. Congress and federal agencies. TIPRO has always given small producers a way to articulate their aspirations and explain their grievances.

TIPRO serves its members through a variety of ways. Its Government Relations Program helps members cut costs and increase revenues by teaming up with a full-time professional staff as they actively lobby at the state and federal level. Through this program, members receive special business alerts informing them of crucial issues requiring their attention.

Other TIPRO services include publications reporting the latest on state and federal regulations, industry news and technology, and membership news. Through its technical services, members can make their businesses more efficient by linking up with the Energy Connection website and can expand their business capabilities by using the Gas Research Institute technology transfer program.

Through TIPRO's marketing opportunities, companies can get name recognition and deliver their message to thousands of oil and gas professionals and through TIPRO's public communications members can assure that their voice is heard through the association's contacts with key journalists in both mass media and the trade press.

Other TIPRO services include troubleshooting for solutions with the association's experienced staff and business networking and educational opportunities at TIPRO's many meetings. Attendance at TIPRO meetings is an excellent way for members to stay in touch.

TEXAS INDEPENDENT PRODUCERS & ROYALTY OWNERS ASSOCIATION

RIG TESTERS, INC.

Rig Testers offices are located at 5333 North Third in Abilene, Texas 1-800-611-3907, 325-673-2771; 3218 Commercial Drive in Midland, Texas 432-570-0421; and 1511 Loving Highway in Graham, Texas 940-549-5681.

Rig Testers, Inc., provides open-hole formation testing services to oil and gas well operators in North, Central, and West Texas as well as in New Mexico. The company conducts business in service shops and offices located in Abilene, Midland, and Graham.

Formation testing involves testing oil and gas wells while drilling rigs are still on location and before completion strings are set. By testing productivity levels, the company is able to advise oil and gas operators on the economic feasibility of setting pipe and attempting completion of the wells.

Rig Testers, Inc., was originally founded in San Angelo in 1963 by Fred Rippetoe (formerly of Johnston Testers) and R.W. Foster (of Foster Testers) in order to service and run formation testing services for area oil and gas operators. In 1970, R. C. Lane joined the company. G. T. (Tom) Myers (formerly of Johnston & Foster Testers) joined the firm in 1974, with R. L (Randy) Myers joining the following year. In 1976, a third shop opened in Graham, expanding the company's fleet to ten trucks in the late 1970s and early '80s. In 1992 the San Angelo shop was relocated to Midland. The original partners brought many years of experience working for large testing companies to their new business.

In 2003 the company celebrated its fortieth anniversary of providing twenty-four hour service, seven days a week, 363 days a year to its customers. In January 2004, Rig Testers, Inc., acquired Permian Testers, Inc., another forty-year-old testing company, founded by Gene Terhune (formerly of Foster Testers) located in Odessa. Don Terhune of that company began as Rig Testers' operations manager while R. L. Myers serves as company president.

Although, Rig Testers, Inc., is not the largest service company it is the largest company, which specializes in Drill Stem Testing in the Texas & New Mexico area, currently running six test trucks. The company's many customers seem to appreciate experienced testers who have a combined testing experience of over 135 years. With tens of thousands of previous test data on file for comparative reviews and utilizing the most modern derivative curve and regression modeling analysis software, the company can obtain the most information possible from a DST.

In more than forty years of providing a valuable service to the oil and gas industry, Rig Testers, Inc. has earned a reputation for conducting business with integrity, efficiency and competence. During this time, the company has merited the respect of its customers, employees and the community it serves. Rig Tester's future business endeavors will no doubt continue to inspire trust in the company's stability and expertise, resulting in its ongoing success.

Risk-taker. Strong-willed. Highly intelligent. Persistent. Innovative. It took a special type of person—someone who exhibited the traits above—to persevere in the oil industry and to be successful in the world of exploration and production. The lives of those people and their exciting world of chance, taking along with the story of how petroleum permeates the core of people's lives unfolds daily at The Petroleum Museum in Midland, Texas. The story of oil is told through art, interactive exhibits, antique equipment, archival collections and narratives.

It took a visionary man who exemplified the description above to dream of a museum to tell the story of petroleum to the world and to ensure it was built. George T. Abell encouraged other petroleum leaders to sign on to his idea, and in 1975, The Petroleum Museum opened its doors. Since then, it has welcomed over one million visitors from around the world.

Every visit to The Petroleum Museum is an opportunity to experience the fun side of learning. The spectacular exhibit wings offer remarkable insight into the scientific and technological world around us, from the age when dinosaurs roamed the Permian Basin to the wild "oil boom" of West Texas to the aerodynamic innovations that changed auto racing, passenger safety, and fuel efficiency forever.

Walk under an ancient sea, stroll the streets of a bustling Boomtown, then literally feel the ground shake beneath you with the force of a nitroglycerin explosion. Marvel at the history captured in the exquisite detail of Prix de West Gold Medalist Tom Lovell's paintings. Inspect the world's largest collection of antique drilling equipment and modern machinery. This interactive museum, located on forty acres, takes you step by step through the dynamic search for black gold.

The Petroleum Hall of Fame gives tribute to more than one hundred men and one woman who have exemplified the tenacious qualities of successful people in the industry. It is dedicated to those who cherished the freedom to dare and whose work and service helped build the Permian Basin.

The Chaparral Gallery enables the Museum to enhance the story of the interdependence of petroleum and surface transportation by telling the pioneering history of the Chaparral Road Racers. The applications of engineering sciences, particularly those in aerodynamics and mechanical engineering, are embedded throughout the exhibits and Chaparral Car displays. Like the oilmen who preceded him, Hall reflected the same qualities in his business designing racecars. He was a risk taker, as well as innovative, strong-willed, highly intelligent, and persistent.

The Petroleum Museum shows that the oil industry has never been one for the faint hearted. Please visit www.petroleummuseum.org for more information.

THE PETROLEUM MUSEUM

Below: Bride's Home at a Wildcat Well by Tom Lovell.

Bottom: Carson's Oilfield Supply Company.

Lancaster Hotel

In the heart of Houston's cultural and business district, almost hidden among the skyscrapers downtown, stands a small luxury hotel that *Conde Nast Traveler* magazine calls "one of the best places to stay in the world." For more than six decades, guests have come to the Lancaster Hotel to enjoy the comfort, elegance and personal service that only this four-star hotel can provide.

Built in 1926, the Auditorium Hotel, as it was known then, was a favorite stopover for noted entertainers in the 1930s. After its basement—now the wine cellar for Charley's 517 Restaurant next door—was converted into a USO entertainment center during World War II, it attracted such celebrities as Helen Hayes, Fay Bainter, and Gene Autry.

The hotel, though, had fallen into disrepair by 1981, when it was acquired by Bill Sharman and a group of investors. During an $18-million renovation, they gutted the interior of the twelve-story structure, converting its 200 rooms into 96, including tensuites, and creating one of the most distinctive hotels in Houston. The Lancaster was designated a national landmark in 1984.

An ever-present doorman greets guests as they arrive at the burgundy canopy-covered entrance of the hotel, just across from the Alley Theater and Jones Hall. As General Manager Sergio Ortiz explains, "We want them to feel pampered from the moment they walk in the door."

That pampering, he says, extends to the amenities, from padded hangers and potpourri bags in every closet to baskets of toiletries and terrycloth robes to fresh flowers, that guests find in every room. And it includes such personal touches as overnight shoe shines, valet service, twenty-four hour room service, daily newspapers at the door, and complimentary limousine service within the central business district.

The decor—the overstuffed chairs, antiques, and eighteenth-century English oil paintings in the lobby and the fine reproduction antiques, imported wallpapers, imported fabrics on windows and beds, and marble bathrooms with brass fixtures in each guest's quarters—remind visitors of the timeless elegance of an English manor. But they also enjoy conveniences that were unheard of during the Auditorium Hotel's heyday: remote-controlled, cable-equipped TVs, video cassette players, hi-fi stereo systems with CD players and mini-bars, all housed in armoires, as well as digital phone systems, Wi-Fi internet, computer data ports, two-line speaker phones with conference call capabilities, and free local and operator-assisted calls.

The Bistro Lancaster features cuisine from Texas, Louisiana, and the Florida Panhandle. In addition, the hotel features three elegant meeting rooms and function space for corporate and social gatherings of 6 to 200 guests.

"Since we're within walking distance of the financial district and the federal and county courthouses, many of our guests during the week are lawyers, bankers and other members of the business community," Ortiz explains. "And on weekends, we cater to theatergoers, performing arts enthusiasts, and, of course, performers because of our location in the Theater District."

Not surprisingly, about half the people who stay at the Lancaster are repeat guests, Ortiz points out. "That's why we keep records on each visitor," he explains. "For example, we'll note if someone likes extra towels or prefers cheesecake with strawberries or doesn't like to be disturbed before noon.

"We pride ourselves on our high standards and our attention to detail," he concludes, "and beyond providing great service, we delight in creating a bond with our guests. We want each guest to feel like a long-awaited houseguest—and to want to come back to the Lancaster, again and again."

The Lancaster Hotel is owned and managed by Houston-based Lancaster Hotels and Resorts. For more information about the Lancaster Hotel, please visit wwww.lancaster.com.

Above: The lobby setting evokes distinguished English manor.

Below: Guest rooms convey a relaxed and timeless elegance.

The publisher and project sponsors would like to express their gratitude for the generous support given to **BLACK GOLD** by the following companies, organizations, and individuals.

Atlantic Communications

Bulldog Specialties, Inc.

E.C. Tool & Supply Company

Harbison-Fischer Manufacturing Company

HEP Oil Company

Holiday Inn Atrium Plaza

Holiday Inn Midland

Thomas D. "Rusty" Howell & Howell Oil & Gas, Inc.

Donald J. Hupp, CPA ITX Corporation

JoJon Petroleum Company

Key Energy Services, Inc.

Magnum Hunter Resources, Inc.

Elton M. Montgomery, Attorney at Law

Panhandle Producers & Royalty Owners Association

John R. Parish

Permian Basin Petroleum Association

Southwest Royalties

Texas Alliance of Energy Producers

Texas Independent Producers & Royalty Owners Association

Texas Oil & Gas Association

TravisWolff

Yuma Exploration & Production Company, Inc.

Pat Long
in memory of
Lawrence R. Hagy

INDEX

About the Author

Dr. Roger Olien

An educator and historian, Dr. Olien has taught courses in business and regional history at the University of Texas-Permian Basin. In addition, he has authored numerous books and articles, many of which have dealt with the history of the oil and gas industries of Texas, including *Wildcatters: Texas Independent Oil Men* and *Oil and Ideology: The Creation of the Cultural Identity of the Petroleum Industry* (both with Diana Davids Olien).

Dr. Olien is also active in petroleum industry public service organizations, and was appointed to the Spindletop 100 Commission by then-Governor George W. Bush, and he sits on the board of directors for the Permian Honor Scholarship Foundation.